# Exploring the History of Hyperbaric Chambers, Atmospheric Diving Suits and Manned Submersibles: the Scientists and Machinery

# Exploring the History of Hyperbaric Chambers, Atmospheric Diving Suits and Manned Submersibles: the Scientists and Machinery

Joseph Stewart

Copyright © 2011 by Joseph Stewart.

| | | |
|---|---|---|
| Library of Congress Control Number: | | 2011901103 |
| ISBN: | Hardcover | 978-1-4568-5723-3 |
| | Softcover | 978-1-4568-5722-6 |
| | Ebook | 978-1-4568-5724-0 |

All rights reserved. No part of this book may be reproduced or transmitted in any form or by any means, electronic or mechanical, including photocopying, recording, or by any information storage and retrieval system, without permission in writing from the copyright owner.

This book was printed in the United States of America.

To order additional copies of this book, contact:
Xlibris Corporation
1-888-795-4274
www.Xlibris.com
Orders@Xlibris.com
92642

# Contents

Introduction ............................................................................................... 9
Photograph Credits ................................................................................. 11
What is Hyperbarics? .............................................................................. 15
UHMS (Undersea and Hyperbaric Medical Society)
    Approved Indications for Hyperbaric Treatment ........................... 18
Ancient Man Under Water ..................................................................... 20
Diving Bells ............................................................................................. 27
1578 William Bourne ............................................................................. 30
1620's Cornelius van Drebbel ................................................................. 31
Robert Boyle 1660's ................................................................................ 32
1634 Mersenne and Fornier ................................................................... 34
1653 De Son ............................................................................................ 36
Henshaw 1662 ........................................................................................ 37
1670's Borelli .......................................................................................... 38
John Lethbridge-Diving Engine 1715 .................................................... 40
Joseph Priestley 1774 .............................................................................. 42
1775 Bushnell ......................................................................................... 44
Antoine Laurent Lavoisier 1779 ............................................................. 46
Beddoes and Watt 1796 .......................................................................... 48
1797 Fulton ............................................................................................. 50
Tabarie 1832 ........................................................................................... 53
1833 De Villeroi ...................................................................................... 54
Junod 1834 ............................................................................................. 55
Charles Gabriel Pravaz 1837 .................................................................. 56
1850-Wilhelm Bauer .............................................................................. 58
1852-Phillips ........................................................................................... 61
1853-Nasmyth ........................................................................................ 63
Phillips-1856 .......................................................................................... 65
Doctor Georg von Liebig 1858 ............................................................... 66
1860 ........................................................................................................ 67
James Leonard Corning 1860 ................................................................. 68
1861-Pioneer .......................................................................................... 70
1861-Alligator ........................................................................................ 71

| | |
|---|---:|
| 1862-David | 72 |
| 1863-Intelligent Whale | 73 |
| 1863-Hunley | 74 |
| 1869-Jules Verne | 76 |
| 1874-Holland | 77 |
| Carlo Forlanini 1875 | 79 |
| Daniel Kelly 1876 | 81 |
| Paul Bert 1877 | 82 |
| Fontaine 1879 | 84 |
| Henry Albert Fleuss-1879 | 86 |
| Carmagnolle-1882 | 87 |
| E.W. Moir 1885 | 88 |
| 1885-Nordenfeldt I | 89 |
| 1885-Goubet | 90 |
| 1897-Lake | 91 |
| William Carey-1891 | 94 |
| Buchanan and Gordon-1894 | 95 |
| 1898-Narval | 96 |
| 1905 John Scott Haldane | 97 |
| 1906 U-boots | 99 |
| 1913 Draegerwerk, Lubeck, Germany | 100 |
| Neufeldt and Kuhnke-1913 | 102 |
| Harry L. Bowdoin-1915 | 104 |
| Campos-1922 | 105 |
| Joseph Salim Peress-1922 | 106 |
| 1927-Momsen Lung | 107 |
| 1928 Dr. Orval J. Cunningham | 108 |
| 1930-Bathysphere | 111 |
| 1930s Siebe Gorman and Company, LTD. | 112 |
| 1939-Rescue Chamber | 114 |
| 1948-Bathyscaphe | 115 |
| 1950's Professor Doctor Ite Boerema | 117 |
| 1955-U.S.S. Nautilus | 119 |
| 1957-Conshelf Habitats | 120 |
| 1964-Sealab | 123 |
| 1964 Draegerwerk, Germany | 125 |
| 1967 Wurzburg, Germany | 126 |
| 1970 Draeger Oxyfulm Monoplace Chamber | 127 |
| JIM Suit-1969 | 129 |
| WASP Suit-Mid 1970'S | 130 |
| 1983 Hyox, Scotland | 131 |

NEWT Suit-1984 .................................................................................. 132
Perry Baromedical Corporation ............................................... 133
Sechrist-Anaheim, California .................................................... 134
HBO Contra-Indicated Items Approved for Dive in a
    Monoplace Chamber ........................................................... 136
Items not approved to be taken into a monoplace chamber ....... 140

Glossary ................................................................................... 143
Bibliography ............................................................................ 149

# Introduction

By definition, hyperbarics could be defined as the science of placing its occupant, whether human or animal, in a structure with pressure greater than that found at sea-level. Hyperbarics used in the medical field for helping humans, entails the use of a chamber with pure pressurized Oxygen. In this book, we will look in to the science behind the discoveries, the history of the very scientists, the quackery, and some truly miraculous healing results. The author chose to write this book, due to the general lack of books on hyperbarics. There are a couple of books containing the history, and a few more purely medical books, but no one book covers the entire hyperbaric scene. This book is in no way meant to be a medical book, although scientific theory and fact will be discussed. This author simply felt that the book would be beneficial to many interested parties in the rapidly-growing hyperbaric field as a brief tutorial to what hyperbarics truly is. After having read this book, the author hopes that even the layman will have a general knowledge of what hyperbarics can do, where hyperbarics came from and how it fits in with our modern medical equipment.

# Photograph Credits

Cover art-courtesy of Draegerwerk Ag and Co.
Cover art-courtesy photo of original drawing by Perot
Alexander the Great in diving bell-courtesy of National Oceanic and Atmospheric Administration
Assyrian frieze-courtesy photo
Gourd breather-courtesy photo
Da Vinci self portrait in red chalk-courtesy of the Royal Library of Turin
Da Vinci diving device-courtesy of the British Library
Sir Francis Bacon-courtesy photo
Diving bell with cannons and barrel-courtesy of the Smithsonian Institute
Breathing tube diver raising cannons-courtesy photo
Edmond Halley-courtesy photo
16th c. diving bell-courtesy photo
William Bourne submarine-courtesy photo
Van Drebbel submarine-courtesy photo
Robert Boyle-courtesy of the Library of Congress
Marin Mersenne-courtesy photo
Rotterdam Boat-courtesy photo
Giovanni Borelli-courtesy photo
1680 Borelli submarine-courtesy photo
Lethbridge diving engine fig.6-courtesy photo
Lethbridge schematic-courtesy photo
Joseph Priestley-courtesy photo
Priestley discovers O2-Joseph Priestley's own drawing
Bushnell Turtle 3 view-Library of Congress prints and Photographs Division; George Grantham Bain Collection
Antoine Lavoisier-courtesy photo
James Watt-courtesy photo

Thomas Beddoes-courtesy photo
1800 Fulton Nautilus-courtesy photo
Robert Fulton-courtesy photo
1833 Villeroi Waterbug-courtesy photo
Charles Pravaz with needles-courtesy photo
Wilhelm Bauer-courtesy photo
Brandtaucher 3 view-courtesy sketch from 1898 German published work
1855 Seeteufel-courtesy photo
Lodner Phillips-courtesy photo
Lodner Phillips 1852 submarine-courtesy photo
James Nasmyth-courtesy photo
Nasmyth Floating Mortar-courtesy photo
James Corning-courtesy photo
1861 Hunley Pioneer-courtesy, sketch probably by McClintock
1861 Alligator-courtesy of the Historique de la Marine; National Oceanic and Atmospheric Administration
1862 Francis Lee David submarine-courtesy photo
CSS Hunley 2-courtesy, drawings published in France based on sketches by William Alexander
1863 CSS Hunley-courtesy of Mobile Daily Herald, July 1902, based on sketch by William Alexander
Jules Verne-courtesy photo
John Holland exiting hatch-courtesy of General Dynamics, Electric Boat Division
Holland submarine-courtesy Department of the U.S. Navy
1875 pneumatic institute-courtesy photo
Carlo Forlanini-courtesy photo
Paul Bert in chamber-courtesy photo
Bert, Paul-courtesy photo
Victorian chamber-courtesy photo
Fleuss apparatus 1879-courtesy Siebe-Gorman
Carmagnolle diving suit-courtesy Vincent Roc Roussey; National Marine Museum, Paris
Caisson air lock-courtesy Scientific American, Nov. 12, 1870
1887 Nordenfeldt III-courtesy photo
1885 Goubet submarine-courtesy photo
Simon Lake-courtesy photo
Lake's patent 1896-courtesy photo
Argonaut I-courtesy photo
Lake Protector-courtesy photo
Campos suit 1922-courtesy Siebe-Gorman

1894 Buchanan and Gordon-courtesy Siebe-Gorman; photo by Alex Hanson
Maxime Laubeuf-courtesy photo
1898 Laubeuf Narval submarine-courtesy photo
Haldane in chamber-Time Life Pictures; Hans Wild
Haldane breathing apparatus 1917-courtesy of Wellcome Library, London
Draeger collapsible chamber-courtesy of Draegerwerk Ag and Co.; Dr. Baixe
Draeger firefighters-courtesy Draegerwerk Ag and Co.
1920s diving suit-courtesy of Neufeldt and Kuhnke
1914 MacDuffy suit-courtesy of Siebe-Gorman
Momsen Lung advertisement-courtesy of the U.S. Navy
Cuningham 1921 chamber-courtesy photo
Steel ball-courtesy of the Dittrich Medical History Center, Case Western Reserve University, Cleveland
Steel ball dining room-courtesy of the Dittrich Medical History Center, Case Western Reserve University, Cleveland
Bathysphere-courtesy of the New York Zoological Society
Diving helmet and dress-courtesy of Siebe-Gorman
McCann rescue chamber cutaway-courtesy of the U.S. Navy Diving Manual
Bathyscaphe Trieste-courtesy of the National Museum of the U.S. Navy
Trieste I-courtesy of the National Museum of the U.S. Navy
Boerema chamber-courtesy of Dr. Ite Boerema, et al
USS Nautilus-courtesy of the Naval Historical Center
Nautilus cutaway-courtesy of General Dynamics Corp., Electric Boat Division
Deep cabin-courtesy of National Geographic, Vol. 125, No. 4, April 1964
Jacques Cousteau-courtesy of the Wyland Foundation
Conshelf II-courtesy of National Geographic
Sealab II-courtesy of the Office of Naval Research
Inside Sealab-courtesy of the Office of Naval Research
Old Draeger monoplace chamber-courtesy of Draegerwerk Ag and Co.
Draeger Duocom chamber-courtesy of Draegerwerk AG and Co.
Draeger diving suit-courtesy of Draegerwerk Ag and Co.
Hyox chamber-courtesy of Divex, Ltd.
NEWT suit-courtesy of International Hard Suits, Inc., Vancouver, B.C.
Sechrist monoplace chamber-author's own photo with thanks to Sechrist Industries
2nd largest chamber-courtesy of Stephen Weir and Aurora St. Luke's Medical Center, Milwaukee, Wisconsin

# What is Hyperbarics?

Hyperbarics may be referred to in many ways. HBO stands for hyperbaric oxygen. H-BOT stands for hyperbaric oxygen therapy. But what is hyperbarics? The simplest definition would be: Hyperbarics is the medical therapy of administering oxygen higher than 21% at a pressure greater than one atmosphere absolute. Most licensed facilities use some variation of this general definition. Modern HBO chambers can administer 100% pure oxygen at 2-3 atmospheres absolute.

But what does atmosphere absolute mean? Imagine that you live at sea level. You live at 1 ATA (1 atmosphere absolute). If you could compress all the gas from the top of your head to the edge of outer space, the pressure of all that gas pushing on the top of your head is called 1 atmosphere. And that gas exerts the same pressure as 33 feet of seawater. So, if a diver were to start at sea level (1 ATA) and dive to a depth of 33 feet into the ocean, that diver would then be at 2 ATA. If he/she dove further to 66 feet deep, the diver would be at 3 ATA.

So, we've deduced that 1 ATM (atmosphere) equals 33 fsw (feet of seawater). Our diver can dive 99 feet under seawater, which would be diving down 3 atmospheres, but when our diver factors in the atmosphere exerted by the air, the diver is now at 4 atmospheres absolute (3+1). Atmospheres absolute (ATA) is most simply put, as the sum of air pressure plus water pressure.

Now for some more complex pressure conversions. 1 atm (atmosphere) equals 14.7 psi (pounds per square inch). That means that the air from the top of your head to the edge of space, OR 33 feet of seawater both exert 14.7 psi on our divers' body. Each foot of seawater equals 0.445 psi.

Pressure can also be measured in millimeters of mercury, just like a sphygomonometer, or blood pressure cuff. 1 atmosphere equals 760 mmHg. In Europe, 1 atmosphere might be referred to as 1 BAR. In America, one might inflate their automobile tires to a pressure of 45 psi. In Germany, Hans might inflate his auto tire to 3 BAR.

Here are some conversions, so that one can do the math him/herself.

1 mmHg = 0.0013 atm
1 fsw = 0.0303 atm
1 atm = 760 mmHg
1 atm = 14.7 psi
1 atm = 33 fsw
1 psi = 0.068 atm
1 fsw = 0.445 psi

Atmospheric pressure is exerted by gases. These gases that we breathe everyday consist of 20.9% oxygen and 78.1% nitrogen. These are usually referred to as 21% O2 and 79% N2. One might notice that there is a missing 1%. This 1% is made up of random gases such as xenon, argon, helium, carbon dioxide, etc.

Before we delve into the scientists and engineers who have made hyperbaric chambers and submarines a reality, let us first review some of the governing bodies that regulate these pressure vessels.

The ASME is the American Society of Mechanical Engineers. The design, manufacturing and testing of any pressure vessel is governed by the ASME. The ASME Boiler and Pressure Vessel Code was formed to protect Americans from being killed by the explosions of faulty boilers. The PVHO (pressure vessels for human occupancy) is the safety standard of the ASME. Any vessel with pressure exerted from the inside or the outside falls under this code. Some exceptions are military submarines, which are scrutinized by their own safety engineers. The NFPA (National Fire Protection Agency) is an organization to protect life and property. The NFPA focuses on preventing fire and electrical hazards in health care facilities. The UHMS (Undersea and Hyperbaric Medical Society) addresses safety in hyperbarics worldwide. The UHMS suggests safe guidelines that may be adopted as codes or standards of safe practice in many countries.

The reason this book is an amalgam of hyperbarics, submersibles, diving suits and diving inventions, is that all of these fields are very closely related. The same science that was used in keeping a man alive in an underwater vessel was used to keep man alive in outer space and in a hyperbaric chamber. Submarines bear tremendous pressures under water and must have breathable gas for the sailors to survive. Other gases are used in the submarine to blow the water out of ballast tanks and to enable the vessel to surface. Jets and spacecraft must be pressurized and must have breathable air for their passengers. Miners, SCUBA divers and bridge-builders must sometimes use hyperbaric chambers to force the accumulated nitrogen out of their bodies. Every human being on earth right now is under some amount of pressure and breathing some oxygen-rich mixture of air. The next time you feel you are under pressure, remember that YOU ARE!

# UHMS (Undersea and Hyperbaric Medical Society) Approved Indications for Hyperbaric Treatment

Air or Gas Embolism-Gas bubbles in the vascular system can actually block the blood flow in vessels. Blocking blood flow in the heart or lungs can be fatal.

CO2 or Cyanide Poisoning-CO2 poisoning can occur due to faulty water heaters, gas leaks and smoke inhalation. Although cyanide poisoning is not considered a great risk, people trapped in fires with burning plastics often fall victim to cyanide poisoning.

Clostridial Myonecrosis (gas gangrene)-The culprit is most often the bacterium Clostridium perfringens, which is an anaerobic bacteria. This means it thrives in areas with little or no oxygen. It is most often treated using antibiotics from the penicillin family along with debridement and hyperbaric oxygen therapy. It is often referred to as gas gangrene because as the bacteria reproduce in the body, it rapidly puts off gas which makes the affected area swell, sometimes to the point that the skin will burst open.

Crush Injury, Compartment Syndrome, Acute Traumatic Ischemia-Ischemia is a restriction in blood flow. Compartment syndrome is when a body part such as the lower leg can be so full of fluid, that the fluid itself can actually restrict blood flow. Crush injuries often result in ischemia, which can cause further complications and tissue death.

Decompression Sickness-Often referred to as 'the bends.' Most commonly found in scuba accidents. Decompression sickness occurs when gas bubbles form in the bloodstream upon decompression, such as a diver surfacing too rapidly from depth. Recompression in a hyperbaric chamber is the best treatment for decompression sickness.

Problematic Wounds-This can include any wound not healing at a normal rate, and often refers to diabetic ulcers of the lower extremities.

Exceptional Blood Loss/Anemia-Hyperbaric oxygen therapy can provide oxygen to most hypoxic areas of the body, even without blood. For the patient who has lost a critical amount of blood, this therapy can help the patient to survive while awaiting a blood transfusion.

Necrotizing Soft Tissue Infections-Also known as necrotizing fasciitis, hemolytic streptococcal gangrene, or flesh-eating bacteria, this disease is often fatal. Hyperbaric oxygen should be used as an adjunctive therapy with antibiotics and surgical debridement.

Refractory Osteomyelitis-A recurring infection of the bone or bone marrow not remedied by antibiotics.

Osteoradionecrosis (radiation tissue damage)-this is necrosis of bone tissue due to the effects of radiation, usually related to the treatment of cancer.

Compromised Skin Grafts and Flaps-Blood flow can be compromised due to radiation or trauma, such as a surgical amputation.

Thermal Burns-Hyperbaric oxygen therapy has been shown to improve the efficacy of antibiotics, reduce scar tissue and increase healing rates in patients with severe burns.

Intracranial Abscess (HBO as adjunctive therapy)-These rare, often fatal abscesses can occur due to tooth infections, sinusitis, head trauma, cranial surgery and meningitis. These abscesses arise from intracranial inflammation, but can be effectively treated with antibiotics, surgical incision and drainage and hyperbaric oxygen therapy.

# Ancient Man Under Water

Since the dawn of time, man has had a relationship with the sea. The ocean has always been a food source, a way of travel, and an undiscovered new world containing mysteries and sea monsters.

As early as 4,500 B.C., breath-hold divers searched the sea floor for food, mother-of-pearl, coral and sponge. There is evidence from 2,500 B.C. of Chinese divers searching the shallows of the ocean floor for pearls and shells.

It is well-known that the ancient Greeks were using the ocean to fill many needs. Fish was a staple of the Greek diet. As found in ancient Greek art, divers harvested sponge and oysters as early as 1,000 B.C. These divers worked naked with only a knife and a net bag to carry their bounty. They coated their bodies with grease to counter the chill of the deep water. The Greeks dropped over the sides of their boats holding a large stone to reach the bottom faster, therefore giving the diver more bottom time.

In a bas-relief dating from 900 B.C., a diver uses an air-filled bladder either for breathing underwater or as an early buoyancy device. This bas-relief is from the palace of King Assur-Nasir-Pal at Ninevah.

Assyrian Frieze 900 B.C.

In 460 B.C., Herodotus told of the Greek diver 'Scyllis,' who salvaged sunken cargo for King Xerxes I. Scyllis was such a successful diver, that Xerxes held him captive onboard a Persian ship. When Scyllis learned that Xerxes was planning an attack on a Greek flotilla, Scyllis grabbed a knife and dove overboard. The Persians searched for him but could not find him and presumed he had drowned. Scyllis returned under cover of darkness, using a hollow reed as a snorkel. He cut all the Persian ships from their moorings and reportedly swam nine miles to return to the Greeks off of Cape Artemesium.

During war, the Greeks often employed divers to cut ships loose from their moorings, so that the ships would drift into rocks and reefs. The Greek diver, possibly using a breathing tube fashioned from reeds, towed bags of provisions to Greek ships under attack or pinned down by ships of the enemy.

breathing from gourd

The philosopher Aristotle made reference to a breathing tube and compared it to an elephants' trunk around 350 B.C. Reed snorkels have been used throughout history by American Indians and Australian aborigines for hunting. Snorkels were used in WWII by Allied troops and also by the Germans during their retreat from the Battle of Kuban. Aristotle tutored Alexander the Great (356-323 B.C.). It is claimed that Alexander the Great was dropped into the sea in a glass barrel to explore what lurked under the water. Alexander the great also made use of a diving bell called *Colimphax* for underwater demolition during the siege of Tyre. Aristotle also made reference to Greek sponge divers using a primitive diving bell. He described this bell as an upside-down kettle filled with air. A diver could logically leave the bell to gather sponge, then briefly return to the bell to take another lungful of air, thus increasing the divers' bottom time.

Alexander the Great in a diving bell

By 168 B.C., commercial divers could be found in all Mediterranean harbors.

It is known that Julius Caesar (102-44B.C.) used divers not only to recover valuable sunken cargo from shipwrecks, but also used divers in military missions much like modern frogmen.

Pliny's book, *Historia Naturalis*, described military divers using a type of snorkel that was held at the surface using floats.

Around 1,000 A.D., a Viking pirate named Otto attacked a clan of Vikings. The defending Vikings sent divers to Otto's ships under cover of darkness to drill holes in the wooden hulls. While Otto could not keep his ships afloat and defend his ships at the same time, Otto and his men were defeated and his ships sank.

During the Crusades (1095-1291 A.D.), divers carried messages to and from the besieged Arabian seaport of Acre. In 1203 A.D., divers were deployed to destroy an underwater blockade protecting an island fort in the Seine River in France.

Around 1300 A.D., Persian divers used goggles with lenses made of polished tortoise-shell.

Leonardo Da Vinci

Artist and inventor Leonardo Da Vinci (1452-1519 A.D.) designed several underwater aids and inventions. One drawing depicts a spiked diving helmet with what was probably a leather breathing tube held above the surface by a buoy device. This helmet was intended to be used by military divers. Da Vinci also designed swim fins for the hands and feet. He drew designs for an armored diving suit that even had a glass facemask. Da Vinci also depicted the use of a leather wine sack or bladder to contain air for breathing under water. Da Vinci made mention of air tanks in his Atlantic Codex, but did not describe them in great detail due to what he called 'bad human nature.' Da Vinci knew his technology would be used by the military only for destruction.

Da Vinci's snorkel

In 1450 A.D., Mariano Taccola described a device for diving that resembled a horses' nosebag. Several hooded diving devices were designed by Vegetius in 1511, Vallo in 1524, Lorena in 1535, Lorini in 1597 and Fludd in 1617.

In 1551 A.D., the Italian mathematician Niccolo Tartaglia invented the diving 'hourglass.' This invention was a type of diving bell that looked like a tall wooden frame with an elongated glass ball in which the diver would stick his head. The diver even had a hand-cranked wench to lower the hourglass into or out of the water.

During a diving exhibition in Toledo, Spain in 1538, two Greek divers went into a diving bell with a lit candle and were lowered into the river. To the crowd's amazement, the diving bell was raised; the divers were dry and the candle was still burning.

Taken from a mid-1500's book on Spanish artillery, an etching depicts a diver wearing a hooded mask with an air tube being held at surface using a buoy. Polished tortoise shell or glass may have been used for the lenses in this mask.

In 1531, Guglielmo de Lorena dove on two of Caligula's sunken galleys using a diving bell of Da Vinci's design.

Famed English scientist Sir Francis Bacon (1561-1626 A.D.) described the use of a diving bell that had legs to hold the bell off the sea floor. This would enable divers to easily leave the bell, work on bottom and return to the bell for a fresh breath of air.

Sir Francis Bacon

In 1578, Englishman William Bourne described the first submarine. This submarine was a two-decked air-tight hull with air pockets for buoyancy. Breathable air was obtained through a tall hollow mast.

In 1616, Franz Kessler designed a diving bell with an internal framework that had benches where the divers could sit and look through small eye ports. Since the bell had a slightly negative buoyancy, the divers were able to walk the bell on the sea bed.

Since the beginning of time, men have always endeavored to explore the depths of the waters. These accounts show that men have always strove to go further and deeper with undersea exploration, taking advantage of the sea's bountiful food supply, hidden resources and sunken riches.

# Diving Bells

The use of diving bells throughout history has been well documented. A diving bell was described in 1665 that was used to recover cannons from a sunken Spanish warship that went down off the Scottish coast in 1585. This bell may or may not have been invented by Scottish philosophy professor George Sinclair or by the German scholar, Sturmius.

Edmund Halley's complex diving bell from the 1690's is well documented. Edmund Halley was an English scientist famed for having discovered the orbit of Halley's Comet. Halley's diving bell stood five feet tall with a five foot base and a three foot top. Several divers could sit on a bench inside the bell. There was also a platform, on which the divers could stand, that hung down three feet below the bottom lip of the bell. The bell was made of heavily-tarred wood and was sheathed in lead to add ballast and to aid in waterproofing the bell. The top portion of the bell was made of glass to provide sunlight for operating. The sides of the bell had viewports and there was a valve near the top to let stale air escape. Lead-covered barrels suspended by ropes were lowered down to the bell. These barrels had a leather hose through the tops and a diver could place the hose inside the bell. When the end of the hose was held higher than the barrel, the pressure would force the air from the barrel to the diving bell.

Halley tested the bell himself with four men to a depth of sixty feet of seawater. The bell was stopped every twelve feet to fill the bell with more air. The reason for this: the deeper into the ocean the bell went, the more pressure the seawater would exert on the air inside the bell, forcing the air into a smaller space at the top of the bell. Adding more air would increase the pressure on the inside of the bell and combat the pressure of the water. To add more air to the bell, a diver would signal topside by pulling a rope tied to bells on the surface ship.

Halley also invented a diving suit to accompany his bell. It consisted of an airtight leather tunic and a helmet with glass viewplates. A leather hose was probably attached to the top of the helmet. Helmets used all the way up into the 1960's used a similar design.

# 1578 William Bourne

One of the earliest detailed descriptions of a submarine was written by innkeeper and science enthusiast William Bourne. His submarine entailed a wooden air-tight two decked ship with large air chambers that could be filled with water using leather plungers. This would decrease the buoyancy, allowing the ship to submerge. He also described a hollow mast that would extend through the water for breathable air.

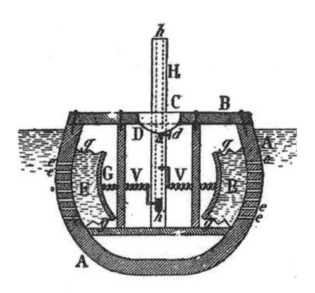

# 1620's Cornelius van Drebbel

Dutchman Cornelius van Drebbel was born in Alkmaar, Holland in 1573. He moved to England in 1604. Van Drebbel was a skilled artist who also studied chemistry, physics, engineering and mathematics. His experiments so impressed King James I, that van Drebbel was appointed the court inventor. Van Drebbel built a wooden submersible that would have appeared as a decked-over rowboat. The wood hull was covered with greased leather to make it watertight. This submersible was propelled by twelve oarsmen and could submerge to fifteen feet of seawater. The vessel used glass viewports. It is thought that van Drebbel had devised a chemical combination that could soak up the $CO_2$ produced by the crew so that the air could be rebreathed, but he kept his recipe secret.

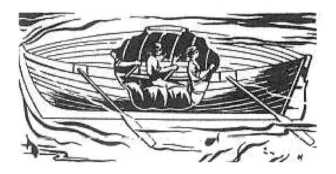

It is claimed that King James I even made a voyage on the submersible and was so impressed, that he ordered two of the subs for military use and requested underwater mines to accompany the subs.

# Robert Boyle 1660's

Robert Boyle was born in Ireland January 25th 1627. He was born into great wealth as the fourteenth child. Boyle had the finest education one could have for the time. He was further educated at Eton, where he studied religion, philosophy, mathematics and the latest advances in physics. Although he spent several years studying under the local parson, he had a fervent passion for science.

Library of Congress

In the 1650's, Boyle had assembled a group of prominent men from all the various fields of science which met weekly in Oxford and London. This group later became known as the Royal Society. By 1680, Robert Boyle was elected society president, but denied taking the position because the required oath violated his religious beliefs.

Boyle was the first scientist to truly publish his work. He meticulously documented his experiments and failures, including his theories, outcomes, parameters of his experiments and his findings. In his first important scientific paper in 1660, *The Spring and Weight of Air*, he discussed using an improved vacuum pump of his own design. He had improved on the cumbersome and inefficient pump of Von Guericke, which required two men to pump vigorously. With Boyle's new pump, vacuum could be sustained with one operator. With the use of his improved pump, he proved that air was a necessity for life. Boyle performed experiments which led him to the discovery of the relationship between pressure and volume of gases. This led to the Boyle-Mariotte Law, which simplified, states that if the temperature is constant, the volume of gas is inversely proportional to the pressure. Boyle went against the theories of basic elements. At that time, it was thought that elements such as salt and water could be broken down no further.

Boyle was a very pious man and died, having never married, in London December 30$^{th}$, 1691. The phrase, 'chemical analysis,' coined by Boyle is commonly used in the science field today.

# 1634 Mersenne and Fornier

Marin Mersenne was a French priest who studied mathematics and physics. Georges Fornier was a naval chaplain, geographer and professor of mathematics. Together, they wrote a detailed book on the building of submersibles, having never actually built a submarine themselves.

Marin Mersenne

They stated that a submersible should be built of metal, thus making it easier to waterproof the vessel. They proposed a visual system for seeing above the waterline. This would enable the sub to navigate or to approach the enemy undetected. They recommended that the vessel be cylindrical in shape and tapered at both ends, enabling the ship to reverse course without turning. This design offers little resistance under water.

They recommended that the ship be armed with a large cannon and be propelled by oars. The ship could have wheels to run along shallow ocean bottoms. Also described were air pumps and an escape hatch. What makes their book more impressive, is that modern submarines are still designed using Mersenne and Forniers' suggestions.

# 1653 De Son

Frenchman De Son built a vessel to be used by the military in the Netherlands. This wooden vessel was known as the 'Rotterdam Boat.' This boat was 72 feet long, 8 feet wide and 12 feet high. The ends of the boat were reinforced with iron and the vessel was designed to ram other ships. This vessel would have ran awash, meaning partially submerged. The propulsion system consisted of a paddlewheel driven by clockwork. Fully wound, the clockwork paddlewheel could run for eight hours on land. However, when placed in seawater, the paddlewheel was not even strong enough to move due to the added resistance.

# Henshaw 1662

In 1662, an English clergyman named Nathaniel Henshaw (1628-1673) invented what could be considered the first hyperbaric chamber. He devised a chamber that could be used to create a hyperbaric or hypobaric environment using organ bellows and valves. The valves could be manipulated to increase or decrease the atmospheric pressure.

Henshaw claimed that acute illnesses such as digestion and respiratory problems could be remedied using hyperbarics. His chamber was called the 'Domicilium.' Henshaw had only theories and no scientific evidence that would have supported his theories. Most hyperbaric scholars agree that the pressures Henshaw achieved would have been very low, possibly in the 1.3 atmosphere range, due to the inefficiency of the equipment he was using. Since oxygen had not yet been isolated as a pure gas, Henshaw was only slightly compressing air, which also contains 79% nitrogen. Oddly enough, Henshaw had no difficulties finding patients to venture inside his chamber.

# 1670's Borelli

Giovanni Alfonso Borelli (1608-1679) was an Italian priest and biologist. He had designed underwater breathing gear, diving bells, and even a submersible. This design is known as the 'Borelli Sub.' This vessel would have been propelled by oars. Depictions of the vessel show goat skins supposedly filled with air for breathing. Additional goat skins were empty and could be flooded with water to allow the boat to submerge. The water could also be forced out of the goat skins using a lever to make the boat more buoyant. Borelli died in 1679. Modern submarines still flood ballast tanks to submerge using his same basic concept.

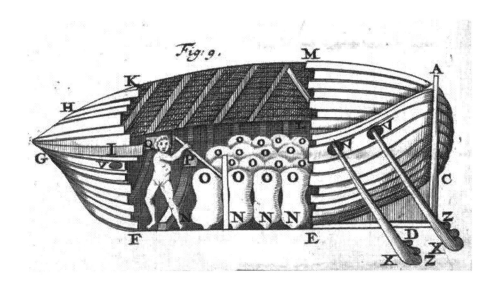

# John Lethbridge- Diving Engine 1715

John Lethbridge conceived of a diving rig that was built similarly to an elongated barrel. The 'diving engine' was approximately six feet long and tapered at the rear. The rig was built of wooden tongue-and-groove slats and reinforced with iron rings. There was a glass porthole over the diver's face and two armholes sealed via oiled leather. The rig was sealed using a 'barrel lid'-style hatch with a gasket fashioned from leather to form an air-tight seal. The entire rig would have been hauled to the surface often to replace the air using a bellows.

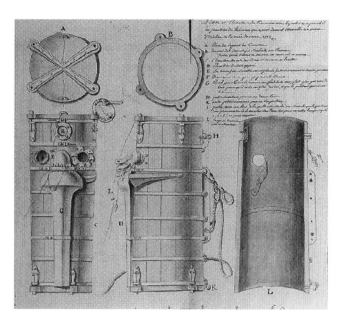

This rig was apparently used successfully to salvage precious cargos from shipwrecks. Lethbridge was employed by the Dutch East India Company in 1724 to salvage silver from a sunken ship in 60 feet of seawater.

# JOSEPH PRIESTLEY 1774

Joseph Priestley was born in Fieldhead, England on March 13th, 1733. He entered grammar school in 1745, learning Greek, Latin, several near-eastern languages, higher mathematics, physics, philosophy and theology. He attended the Dissenting Academy in Daventry for four years. Priestley became a minister in 1755. He then went on to teach higher sciences at the Academy of Warrington. In 1765, Priestley received a doctorate of laws from the University of Edinburgh for his *Chart of Biography*. He was received by the Royal Society in 1766. It was in Warrington that Priestley met and became friends with Benjamin Franklin. Priestley wrote a paper entitled *The History and Present*

*State of Electricity, with Original Experiments* in 1767. This paper described some of his experiments with electricity. Priestley became a librarian at Paris in 1770, which at that time, was a very prestigious posting. In 1772, he was elected to the French Academie of Sciences. Priestley published another paper called *The History of the Present State of the Discoveries Relating to Vision, Light and Colours*. In 1774, he wrote *Experiments and Observations on Different Kinds of Air*, in which he detailed his discovery of oxygen together with eight other gases such as chlorine and carbon monoxide. He refined oxygen by heating mercuric chloride and referred to oxygen as 'dephlogisticated air.' A 'phlogiston' was the term used at that time for a weightless principle of flammability. Priestley was elected to the St. Petersburg Academy in 1780. Around this time, both the American and French revolutions were in full swing. Priestley openly expressed that he thought the French Revolution would bring about the biblical apocalypse. Due to his open beliefs, an angry mob destroyed his home and laboratory in Birmingham, England. This act influenced Priestley to move his family to Northumberland, Pennsylvania in 1794, where he established a new home and laboratory on the Susquehanna River. He became close friends with Thomas Jefferson and John Adams. Thomas Jefferson is quoted as having said, "(Priestley is) . . . one of the few lives precious to mankind."

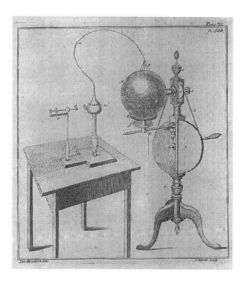

Priestley discovers oxygen

There Priestley died on February 6th, 1804. Along with discovering oxygen, Priestley improved on the form of shorthand writing, invented carbonated water, and helped man to better understand respiration and photosynthesis.

# 1775 Bushnell

David Bushnell was born in Saybrook, Connecticut around 1742. He attended Yale from 1771 to 1775. While at Yale, one of Bushnell's professors stated that gun powder could not explode underwater. Bushnell, doubting this theory, fashioned a watertight charge of gunpowder and a firing mechanism. Bushnell found that the blast force was magnified underwater. This discovery inspired Bushnell to consider how an underwater mine could work. But Bushnell would need a way to place these mines covertly. Since a ship or a swimmer could be easily detected, Bushnell considered underwater travel. Bushnell then designed the first submarine to be used for war.

At this time in history, the relationship between America and the British was strained. There were rebellions and rumors of war. Bushnell built what was to become known as the "American Turtle." Although no exact description was ever written, much is known about the design. This was a one-man craft, approximately six feet high. This sub had a lead ballast at the bottom and an air supply to remain submerged for thirty minutes. The Turtle had two vent tubes through the top of the craft that were covered with flaps to keep the water out. This sub is said to have had viewports. The pilot entered the vessel through a top hatch and sat on a bicycle-like seat. The sub was probably controlled by foot pedals. There were two propellers; one for vertical movement and one for horizontal movement. There was a rudder and two bilge pumps. One foot pedal allowed water into a ballast tank to allow the boat to submerge. Bushnell's instruments included a compass, enabling him to take bearings of the enemy ships before submerging. Bushnell also designed a pressure-operated depth gauge. His instruments were dimly lit with phosphorous paint.

This submarine was designed to attach an underwater mine containing 150 pounds of gunpowder to the hull of enemy ships. The mine would have been hung on a large screw by a rope and would have had a clockwork detonation device. George Washington saw the potential in Bushnell's creation and allotted him money and munitions for use toward making the Turtle successful.

As Bushnell was not physically up to the challenge of piloting the sub, army sergeant Ezra Lee was trained. Soon, Lee was given the mission of destroying the British ship HMS Eagle. Ezra Lee tried desperately to anchor the screw of the mine into the Eagle's hull, but struck metal. Many ships of that era had copper plating or iron banding on the hull. His air running out rapidly, Lee surfaced. He was soon spotted by the British. The alarm was sounded. Lee submerged as fast as he could and managed to escape. The British ship was unharmed.

Later that year, the Turtle was being transported onboard a surface ship and was sunk by cannon fire. Bushnell managed to recover the Turtle, but it saw no further action. Bushnell later became an army engineering officer, and after the war, even commanded the army Corps of Engineers at West Point. After Bushnell resigned, he practiced medicine until his death in 1826.

# Antoine Laurent Lavoisier
# 1779

Antoine Lavoisier was born on August 26th, 1743 in to French nobility. Due to the death of his mother, the Parisian inherited a massive fortune at only five years of age. Lavoisier attended College Mazarin from 1754 to 1761. There, he studied the latest in science and mathematics. By the age of 25, Lavoisier was elected to the French Academie of Sciences, due in great deal to his essay on street-lamps. In 1771, Lavoisier being 28 years old, married Marie-Anne Pierette Paulze, who was then thirteen years old. He became chairman of the board of the bank that would later be known as the Banque de France. Lavoisier also graduated with a law degree, however, never practiced law. Although Lavoisier's accomplishments to the sciences are great, it should be noted that he was known for taking the work of other scientists of the time and expounding on their works.

In 1769, Lavoisier played a role in creating the first geological map of France. He went on to name hydrogen, Greek for 'water-former,' after discovering that when hydrogen bonds to oxygen, water is created. Again, Lavoisier was expounding on Priestley's theories. Lavoisier experimented with oxygen's involvement in plant and animal respiration. He discovered oxygen was needed to form rust. By 1779, he had named oxygen, Greek for 'becoming sharp,' due to the acrid taste of oxygenated acids. Lavoisier determined that air was actually a mixture of independent gases. In 1789, Lavoisier wrote *Elementary Treatise of Chemistry*, considered by many to be the first true chemistry textbook. This textbook contained his law of conservation of mass, which simplified, states that even though matter may change form, the mass is not lost. This book demonstrated that water, thought to be an element at that time, could be further broken down into two elements of hydrogen and oxygen.

Among Lavoisier's other accomplishments, he used a calorimeter to measure the heat generated from the burning of certain gases. He discovered that combustion requires oxygen. Lavoisier is also credited as having compiled the first list of chemical elements.

During the Reign of Terror, Lavoisier was accused by Jean-Paul Marat of blackmail and selling watered-down tobacco. He was found guilty and beheaded at the guillotine in Paris on May 8$^{th}$, 1794. Ironically, Lavoisier was exonerated by the French government shortly after his execution.

# Beddoes and Watt 1796

James Watt                    Thomas Beddoes

Although the famed Scottish inventor James Watt is well-known for perfecting the steam engine, he is lesser-known for his work with Thomas Beddoes in oxygen therapy. Watt's improvements to the steam engine were pivotal in bringing about the industrial revolution. Watt was a member of the Lunar Society, the French Academy of Sciences, the Royal Society of Edinburgh and the Royal Society of London.

Beddoes formed the Pneumatic Institute in Hotwells, England. He hired a brilliant young man named Humphrey Davy to run his institution.

At this institution, oxygen treatments were given to patients free-of-charge for several conditions including asthma, palsy, dropsy (edema), venereal complaints, scrofula, the 'king's evil' (lymphatic tuberculosis) and consumption (tuberculosis). Beddoes and Watt never claimed that the oxygen would cure the patient. They were simply testing their theories that adjusting the oxygen percentages inspired could be beneficial. However, since they could probably only raise the oxygen percentage to about 30% at most, and normal air contains 21% oxygen, the benefits would not have been noticed.

Beddoes and Watt made plates, or blueprints, of their medical devices so that others could reproduce and take advantage of their inventions. Beddoes and Watt also designed corrugated breathing tubes and mouthpieces for breathing oxygen. Variations of their basic designs are still used today.

# 1797 Fulton

Robert Fulton was born in Lancaster, Pennsylvania in 1765. He went on to England, where he struggled to earn a living as an artist. He was already known for his work with steamships and returned to designing mechanical equipment and canal systems. In 1796, Fulton went to France to promote his idea for a submarine that he boastfully claimed could destroy the British fleet.

Fulton wrote the French government offering to build a 'mechanical nautilus.' The French government soon appointed a panel of French scientists to decide if the project was feasible. Fulton created a clockwork model for a submarine that would use the same type of mine as Bushnell. This model submarine had a cigar-shaped hull with a conning tower and vent tubes. It had a four-bladed screw propeller that was to be hand driven by two or three sailors. Fulton would also make use of Bushnell's depth gauge. Fulton proposed using dive planes for vertical movement, but switched to using a vertical propeller, as the scientists thought the vessel would not move quickly enough through the water to make the dive plane work. Another interesting feature of Fulton's design was a collapsible sail on a collapsible mast that could be used while the ship was at surface. This sail and mast folded down and stowed neatly in a groove on the deck of the sub.

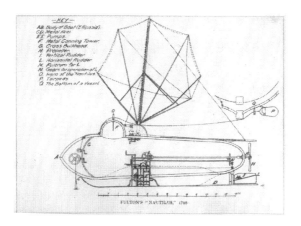

Fulton's Nautilus

However, the French government did not pursue Fulton's submarine idea. Fulton, becoming disenchanted, intentionally leaked information to the British, in hopes that the British government would return information to France. Fulton hoped that British fear would encourage the French to build the weapon. The British did hear of Fulton's design, but did nothing.

At this time, Fulton had no income since the French were not buying his design. He then designed and patented the 'panorama.' This was a large picture painted on the inside of a dome. He erected a building in Paris for the panorama, which was a depiction of Moscow in flames. The panorama was in instant success and Fulton became wealthy. With his new-found wealth, Fulton decided to build his own submarine. This ship was 21 feet long with a seven-foot diameter. The framework was iron and the hull planking was sheathed in copper. Fulton later added a spherical copper tank to hold compressed breathing air. He used candles to light the instruments. The ship had its first trial in the Seine River in 1800. Fulton submerged with two sailors to the depth of 25 feet for 45 minutes. The submarine traveled somewhere between two and four miles per hour while submerged.

Fulton finally received French backing and soon went to destroy two British ships that were part of a blockade at Le Havre. As soon as Fulton entered the theater of war, the British ships hoisted sails and moved out of range. British spies had already alerted the ship commanders of Fulton's attack.

Fulton later tested the Nautilus against a French sloop. With military officials in attendance, Fulton blew the sloop out of the water. However, the French aristocracy was disgusted by the display and thought the treacherous attack to be uncivilized. They wanted nothing to do with such an invention. Fulton then went directly to Napoleon, who had shown some interest in the sub, but since Fulton wanted payment in advance, Napoleon also shied away from Fulton's invention.

Fulton then went to England, asking for their patronage. The British assigned Fulton to design mines and torpedoes, but only to keep him from designing weapons for the French. Soon after, Horatio Nelson crushed the French fleet at the Battle of Trafalgar. The British let Fulton go, as he was no longer considered of any use or threat.

Fulton returned to America, where he tried in vain to interest President Jefferson in his submarine. Again, Fulton was unsuccessful. Fulton will forever be remembered for his submarine, the 'Nautilus.' Nautilus was used as the name of Captain Nemo's ship in the 1870 Jules Verne classic, *20,000 Leagues Under the Sea*. Nautilus was again used by the U.S. Navy in 1954 when the first ever nuclear submarine was also christened the 'Nautilus.'

# Tabarie 1832

In 1832, Emile Tabarie of Montpellier, France, theorized that pressurized oxygen would be beneficial for any number of illnesses. Tabarie believed that increasing the pressure gradually would result in safer compressions. He also had the capability to maintain specific pressures within the chamber. The treatments usually lasted two hours. Today, modern chambers compress slowly, usually between 2 and 5 psi per minute. Modern chambers also maintain a constant pressure and the treatments usually last about two hours. All of these features, still in use today, contribute to safety in hyperbaric medicine.

# 1833 De Villeroi

In 1833, Brutus De Villeroi built what came to be known as the 'Waterbug.' This boat was 10 ½ feet long with a two-foot diameter. It was designed to be operated by a three-man crew that would use duck-like paddles for propulsion. De Villeroi tried unsuccessfully to pedal his invention to the French and Dutch. De Villeroi may have referred to his submarine as 'nautilus' as an homage to Fulton. Another interesting note is that young Jules Verne attended college while De Villeroi was employed there as a professor.

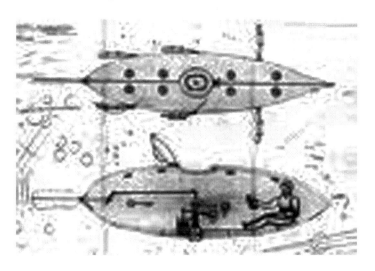

De Villeroi's Waterbug

# Junod 1834

In 1834, a French doctor named Junod built a spherical copper hyperbaric chamber based on the designs of James Watt. Junod theorized that the chamber could treat most pulmonary ailments. The chamber was capable of reaching pressures of 4 ATA, which is equivalent to being 100 feet under seawater. Pressure was achieved by using a manual pump. However, Junod compressed the patients rapidly which caused complications, and many scientists ridiculed the technology. Oddly, many patients claimed to have felt euphoric and relaxed after the treatments, probably due to the narcotic effect of nitrogen at pressure. Junod is credited with having made the first hyperbaric chamber.

# Charles Gabriel Pravaz 1837

Charles Gabriel Pravaz, born 1791, was a noted French orthopedic surgeon who is renowned for his invention, or perfection, of the hyperbaric needle. He developed a screw-operated hollow silver graduated hypodermic version of the needle. This needle allowed several conditions, namely pain, to be expediently treated.

In 1837, Pravaz developed the largest hyperbaric chamber of its time. Regrettably, little is known about the actual chamber. Pravaz treated several ailments including cholera, deafness, rickets, pertussis, laryngitis, tuberculosis, menorrhagia, conjunctivitis and several others. His chamber could reach pressures of 2 to 4 atmospheres absolute. These treatments were referred to as 'compressed air baths.' His treatments attracted patients from all over Europe and even America. Pravaz inspired the creation of 'pneumatic centers' that were built in several of the major European capital cities.

Among his other achievements in the field of orthopedic surgery was the invention of a wire mesh system that could maintain bones in their correct positions. Pravaz died in June, 1853.

# 1850-Wilhelm Bauer

War had broken out between Germany and Denmark in 1848. A Bavarian cavalry corporal named Wilhelm Bauer worked with a shipbuilder to create his submarine design, the 'Brandtaucher,' which translates to 'burn diver.' It was an awkward craft that looked like a box flattened at each end. This boat was made of riveted sheet iron. It was propelled by two men on a treadmill that was attached to a propeller. A third crewman steered the vessel. Buoyancy was achieved with ballast tanks and trim was adjusted by screwing a heavy weight along a threaded iron rod moved by turning gears. This sub had two large square windows on each side but did not have a conning tower.

# Exploring the History of Hyperbaric Chambers, Atmospheric Diving Suits and Manned Submersibles: the Scientists and Machinery

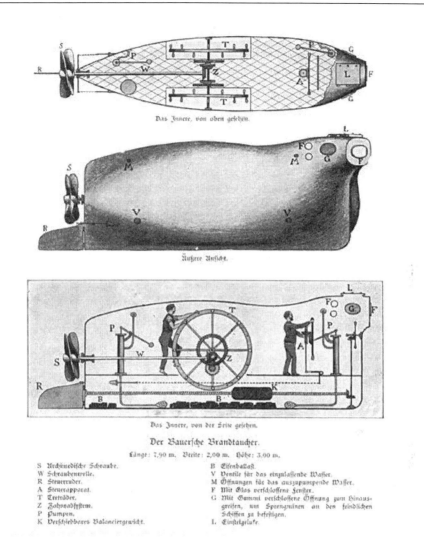

## The Brandtaucher

On the Brandtaucher's second mission, the weights used to adjust trim slid too far forward and the boat went into a steep dive. The submarine was stuck on the ocean bottom in sixty feet of seawater and taking on water. Due to the pressure of nearly six tons pushing against the hatch, the men could not escape. Bauer ordered his men to flood the submarine. The pressure of the incoming seawater compressed the remaining air in the cabin so greatly, that after five hours, the hatch blew open and the crew were shot up to the surface.

Seeteufel

The Russians paid Bauer to build a larger submarine called the "Diable Marin," meaning sea devil. This submarine would have been 52 feet long with a 12 foot diameter. This submarine was planned to have heavy rubber gloves in the bow, so the crew could attach 500 pound bombs to enemy ships. However, the fighting ended before the ship was completed.

# 1852-Phillips

Lodner Darvontis Phillips (1825-1869) was a shoemaker from Indiana who had an interest in underwater engineering. He is known for building two submarines. The first submarine collapsed at twenty feet of seawater. The second, however, reached a speed of four knots and was tested to a depth of one hundred feet. The U.S. Navy showed no interest in his designs, so Phillips sold the second sub to a private owner. The owner and his dog were found drowned in the submarine and Phillips decided never to build another submarine. Lodner Phillips also patented a steerable submarine propeller and in 1856 patented an atmospheric diving suit.

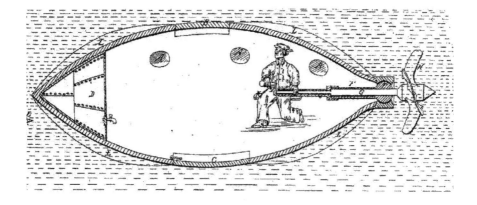

Phillips' steerable propeller

# 1853-Nasmyth

A Scottish inventor named James Nasmyth, who was well known for his work with steam-powered machines, built the first power-driven submarine in 1853. This boat actually ran awash and was named the 'Floating Mortar.' However, newspapers dubbed it the 'waterhog.' This boat was 80 feet long and 30 feet wide, with 10 foot thick walls made of poplar wood. This boat ran under steam power and had smoke stacks and a small conning tower. Since this ship was designed as a battering ram, the nose was reinforced with heavy brass and contained a mortar that could fire underwater. This submarine was never used.

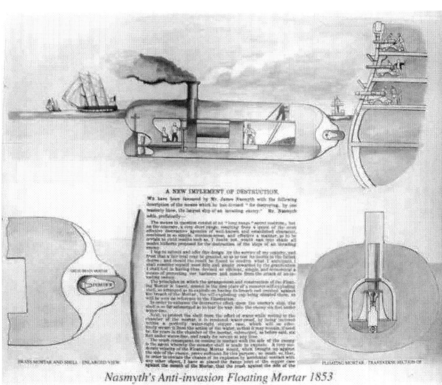

*Nasmyth's Anti-invasion Floating Mortar 1853*

# Phillips – 1856

An American named Lodner Phillips designed a 'diving armour' that consisted of a main body cylinder with arms and legs attached by ball-and-socket joints. The arms had hand-controlled pincers and there was a porthole over the face. The hatch would be sealed at the top of the suit and a heavy chain was attached to the top of the rig to raise or lower the suit into the water. An umbilical hose, probably made of leather or rubber was used to supply air to the rig. This rig was also designed with a hand-cranked propeller, but the rig was probably never actually built.

# Doctor Georg von Liebig
# 1858

Born in Munich February 17th, 1827, Georg von Liebig was the son of famed chemist Justus von Liebig. After studying in Giessen and Berlin, von Liebig went on to work for the English-Indian Company and became a professor at Calcutta. Von Liebig worked as a doctor of saline medicine at Bad Reichenhall, Germany in 1858.

There, von Liebig researched methods of using brine aerosols and pine vapor to treat pulmonary disorders. He is responsible for bringing 'pneumatic' or hyperbaric chambers to Bad Reichenhall.

The first chamber to be constructed for Bad Reichenhall was formed by six cylinders which were riveted together. The construction was steel with rectangular windows which could be used to observe the patients. This bulky structure was over nine feet tall.

As the spa developed, five chambers of steel and concrete were used. Up to 64 patients could be treated simultaneously. Some of the chambers even had furniture of wood and wicker, telephones, reading material and even running water.

Although hundreds of patients were treated and von Liebig discovered many ways to improve safety, several patients suffered from effects such as carbon monoxide poisoning, swelling of the joints, heart damage and ruptured tympanic membranes.

Von Liebig discovered that if the pressure was increased gradually, less side- effects presented. He even preconditioned new patients with 'shallow dives,' or low- pressure treatments. Von Liebig improved the ventilation so that carbon dioxide levels would remain within normal limits. Fresh air was exchanged often in the chambers. Today there exists the Georg von Liebig Hospital, a large medical center in Bad Reichenhall, Germany.

# 1860

The first hyperbaric chamber in North America was built in Oshawa, Ontario, Canada. It was later shipped to New York to treat nervous conditions.

# JAMES LEONARD CORNING 1860

James Leonard Corning was born in Connecticut in 1855. He graduated from the University of Wuerzburg, Germany in 1878. He later went on to become a neurologist in New York and brought the first hyperbaric chamber to America in 1860.

This chamber was approximately six feet in diameter and was the first to employ an electric air compressor.

As a neurologist, Corning observed severe decompression sickness at the Hudson River tunnel site. As the workers would finish their day working under water level, they would suffer terrible muscle pain and even paralysis. Corning associated these symptoms with disorders of the spinal chord and used hyperbarics to treat the workers. The treatments usually lasted one to two hours at pressures up to three atmospheres absolute.

Corning later invented spinal anesthesia by injecting cocaine into the area around the spinal chord in 1885, thus changing surgical medicine forever. James Corning died in 1923.

# 1861-Pioneer

In 1861, a submarine was designed and created by James McClintock. This boat, the 'Pioneer,' was operated by naval captain and cotton merchant Horace L Hunley. This ship was twenty feet long and utilized a three-man crew; two men to crank the propeller and one man to steer. This ship performed a demonstration in Lake Pontchartrain in 1862, where it sank a barge by towing a floating torpedo. Just one month after the demonstration, the U.S. Navy overtook New Orleans and the confederate Pioneer was scuttled by its crew. The Pioneer was eventually sold for scrap in 1868.

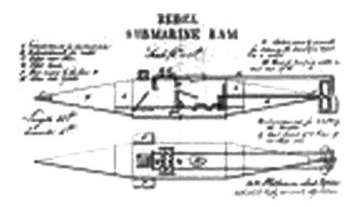

Hunley Pioneer

# 1861-Alligator

In 1861, Brutus de Villeroi got a contract from the U.S. Navy to build the 46-foot long 'Alligator' for the sum of $14,000. It was to be powered by 16 oarsmen, but was modified to use a three-foot diameter hand-cranked propeller. This boat had onboard air compressors and an air-scrubbing system and became the first U.S. navy submarine, although it was never truly designated by the Navy. The Alligator sank while being towed during a storm in 1863.

DE VILLEROI'S SUBMARINE BOAT, SEIZED BY THE GOVERNMENT AT PHILADELPHIA, MAY 16TH, 1861.—FROM A SKETCH BY OUR SPECIAL ARTIST.

# 1862 - David

Confederate captain Francis D. Lee created a small submarine called 'David' in 1862. This boat was so named after the Bible story of David and Goliath. David was smaller than Goliath but brought the giant down with a sling and stone. This submarine was designed to push a spar torpedo into enemy ships. A spar torpedo consists of high explosives on the end of a long pole. The boat was built by the Southern Torpedo Boat Company in Charlestown, South Carolina.

For its first mission, the David was to attack the Union ship 'New Ironsides.' The David was fifty feet long with a nine foot diameter. It ran awash using a steam engine. As it neared its target, New Ironsides could not aim her cannons low enough towards the water to hit the David, so the Union sailors used their rifles. The David's spar torpedo exploded against the hull of New Ironsides and the Union ship rolled practically on its side, but then righted herself. However, the David was blown several feet in the air and smashed back down in the water. Two members of the David's crew managed to reboard the vessel and the David escaped. The remaining crewmembers were taken prisoner.

# 1863-Intelligent Whale

A group of northern investors formed the American Submarine Company in New Jersey. They worked on the building of the submarine, 'Intelligent Whale,' for nine years. This sub was 26 feet long with a crew of thirteen men. It was hand-propelled and had an air-lock for divers to exit the boat while submerged.

The owner, O.S. Halstead tried to convince the U.S. Navy to buy the vessel, so trials were held in 1872. The Whale failed miserably. Halstead was eventually murdered, probably by the lover of his mistress. The Intelligent Whale sank often and had to be brought back to the surface. The Whale claimed the lives of 39 sailors in all. The Intelligent Whale is now on display at the Militia Museum in New Jersey.

# 1863-Hunley

Horace L. Hunley, the cotton merchant and naval captain, also built 'David' type submarines. His boats were 25 feet long with a thin cigar-like shape. The Hunley Davids were hand-propelled by eight men. This ship was designed to run awash until the sub was nearly on top of its target, as the ship had no reserve air supply. This boat was far from being successful. It sank several times before it even went into action. It was once sunk by the wake of a passing steamboat. Another time, it sank while anchored. In 1864, its mission was to destroy the Union ship 'Housatonic.' The Hunley David rammed its torpedo into the Housatonic, but the detonation was premature and the Housatonic was blown out of the water, along with the David. In total, the David had drowned 35 sailors. The CSS Hunley, as this David became known, was recovered in 2000 and is now being preserved for display.

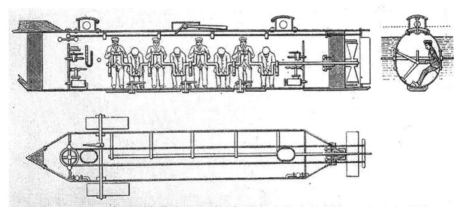

Fig. 175 à 177. — Le *David* de Hunley reconstitué d'après les dessins de M. William-A. Alexander (1863).

# 1869-Jules Verne

In France in 1869, Jules Verne wrote his novel, *Twenty Thousand Leagues Under the Sea*. The book became a bestseller. The name of the mysterious submarine in the book was the 'Nautilus,' probably taken from Fulton's invention. In this book, Captain Nemo piloted a high-speed submarine driven by an engine with a secret fuel source. The crew used scuba gear and gathered all they needed to survive from the sea. This book did much to inspire a new passion for deep sea exploration.

# 1874 - Holland

John Phillip Holland was born in Ireland in 1841. He immigrated to America and became a school teacher in New Jersey. Holland designed a submarine that was 15.5 feet long, three feet wide and 2.5 feet high. This submarine had a foot-controlled propeller and ballast tanks to submerge.

When the U.S. Navy showed no interest in his boat, Holland looked to the Fenians, a group of Irish revolutionaries. In 1878, the Fenians gave Holland their patronage and Holland built the 'Holland I' using a gasoline engine. The Fenians were impressed by the success of the Holland I and ordered a larger boat from Holland.

In 1881, the 'Fenian Ram' was built. This boat had an air-powered cannon and could achieve a speed of nine knots. It could remain submerged for an hour at a depth of sixty feet of seawater. However, the Fenians became frustrated with Holland's delays and stole the boat. The Fenians hid the boat in a shed in New Haven, Connecticut, where it remained forgotten for 35 years. The submarine was donated to the city of Patterson and is now on display in West Side Park.

In 1883, Holland went on to form the Nautilus Submarine Boat Company. In 1897, Holland built the 'Holland VI.' This boat was 54 feet long, using gasoline and electric powered engines. It could operate forty hours and achieve seven knots an hour. It had a crew of six, an air-launched cannon and could carry three torpedoes. It even had a toilet.

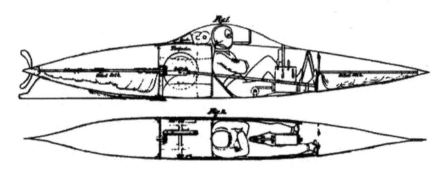

On April 11th, 1900, the U.S. Navy purchased the Holland VI for $150,000. The Navy eventually ordered seven more Hollands, but at the time, Holland was having financial trouble and sold the Holland Torpedo Boat Company to the newly formed Electric Boat Company. The Electric Boat Company became the largest manufacturer of submarines in the world and went on to build the first nuclear-powered submarine called the 'Nautilus' in 1955.

# Carlo Forlanini 1875

Carlo Forlanini was born in Milan, Italy June 11th, 1847. He received a doctorate from the University of Pavia in 1870. Forlanini specialized in the treatment of tuberculosis and other respiratory disorders. He went on to teach at his alma mater, the University of Pavia, and the University of Turin.

In 1875, Forlanini started a 'pneumatic institute' in Milan, Italy. He is credited with making the first horizontal chamber, which was composed of riveted steel. The cylindrical chamber was approximately ten feet tall and fifteen feet long. The chambers were able to be lighted by using external lamps just outside of the rounded windows. The interiors of the chambers were lavishly decorated in baroque fashion with ornamental mirrors, Persian rugs and even furniture.

Carlo Forlanini went on to invent the artificial pneumothorax. He used a surgical needle to create an opening to allow air in or out of the lung. This procedure is still used today. Forlanini died May 26th, 1918. the Carlo Forlanini Institute in Rome is named in his honor.

# Daniel Kelly 1876

Daniel Kelly filed for a patent on his improvements to the compressed air bath in 1876. His chamber was made of wrought-iron that was riveted at the seams. The chamber was of the vertical cylinder design. It had an inward-opening door with rubber gaskets around the door jam to maintain pressure. The outside of the door additionally had six latch-hooks. The chamber was designed with the capability of compressing more than one patient at a time. The patients would have sat upright in chairs. This design entailed pressure gauges and safety valves. The chamber had one window on the side of the chamber and one on the very top allowing light to shine in. The most interesting aspect of Kelly's design was an air-lock. He described the air-lock as a place to pass items such as food through to the patient without losing pressure.

# Paul Bert 1877

BERT (PAUL)

Paul Bert, a zoologist and politician, was born in Auxerre, France December 11th, 1833. He went on to become a professor at the Sorbonne and was elected to the French Assembly in 1874. Bert also became a member of the Academy of Sciences.

Bert is most renowned for studying the effects of pressure. His chamber consisted of two cylinders that were riveted together. This chamber was capable of producing hyperbaric and hypobaric conditions. Bert used a caged bird in the chamber to detect toxic gases. This technique is still used in coal mines of third-world countries today.

Paul Bert also studied the effects of pure oxygen on organisms. For his testing, he often used animals in a smaller separate chamber. This is where he discovered oxygen toxicity. Oxygen toxicity is a physiologic response to high levels of oxygen. This can affect several systems causing seizures, pulmonary complications and visual changes.

Bert discovered nitrogen narcosis, which is still often referred to as the 'Paul Bert effect.' When an organism under pressure breathing partial nitrogen decompresses too rapidly, nitrogen bubbles are formed in the blood stream. This can cause symptoms ranging from muscle pain and confusion to severe issues such as cardio-pulmonary symptoms or embolisms in the blood stream.

Paul Bert (padre de la fisiología hiperbárica) en una cámara experimental.

Bert found that altitude sickness is due to the lack of oxygen at higher elevations. Paul Bert's book, which translated to English is entitled, *Barometric Pressure: Researches in Experimental Physiology* was even used to develop aviation medicine practices used in World War II.

In 1886, Bert was appointed Governor-general to what is now Viet-Nam. This area had several French colonists at that time. Bert died of dysentery in Hanoi December 11th, 1886.

# Fontaine 1879

Fontaine was a French surgeon who developed the first mobile hyperbaric chamber with a matching mobile lever-pump. It was a large chamber comprised of riveted steel plates and was sizable enough to treat up to twelve patients at once. It also had a small air lock and possibly up to ten porthole windows. This unit had what could be considered an air tank for storing the air compressed by the lever-pump.

Fontaine realized that nitrous oxide was more effective during pressurized surgeries. He also noted that there were fewer side effects of nausea and cyanosis after surgeries performed in the chamber. Fontaine actually set the chamber up as a mobile hyperbaric surgical suite and treated several patients. At this time, hyperbaric chambers were already in all the major European cities and the success of his surgeries along with faster recovery times was gaining notice.

Unfortunately, Fontaine was killed during the construction of another hyperbaric chamber at a pneumatic institute.

# Henry Albert Fleuss - 1879

Henry Fleuss was an English merchant marine officer. He invented the Fleuss vacuum pump. It resembled the Guericke pump, but maintained more continuous suction. In 1878, Fleuss invented a rebreather that consisted of a rubber mask that led to a breathing bag. The bag was fed by a copper oxygen tank. The oxygen would have been regulated manually. The exhaled CO2 was scrubbed by passing through rope yarn soaked in caustic potash.

Fleuss marketed his invention with the Siebe-Gorman Company where improvements were made to his invention. This rebreather became known as the Davis Lung or Momsen Lung and was later used in WWII and mining safety. Fleuss later became a master diver for the Siebe-Gorman Company.

# Carmagnolle-1882

Two French brothers designed a rig with several articulations and 25 small portholes. This suit, appearing like a robot from an old science-fiction movie, was actually constructed, but may have never been tested. It is on display at the Musee Oceanographic in Monaco.

# E.W. Moir 1885

E.W. Moir was a brilliant British architect, engineer and inventor. He was the first to employ medical air locks or recompression chambers for the use of caisson workers.

Caissons are deep shafts below riverbeds or even sea beds used in the building of tunnels and bridges. Moir found that during the construction of the Hudson tunnel, 25% of the workers were dying every year. By employing the medical air lock, or 'hospital lock,' Moir reduced the death rate to 1.6%.

Caisson air lock entrance

Moir also consequently improved the working conditions of the 'sand hogs' not only by insisting on the use of the air locks, but also by ensuring that the workers had warm clothes and blankets and hot coffee upon exiting the shafts. He also realized that there was a correlation between alcohol consumption and increased risk of decompression illness.

# 1885-Nordenfeldt I

In 1885, the 'Nordenfeldt I' was built. This boat was 64 feet long and carried one external torpedo. It used steam power on the surface and stored-steam power while submerged.

In 1887, the 'Nordenfeldt III' was built. This submarine was 123 feet long and could operate at 100 feet of seawater. It was purchased by the Russians, but ran aground before the Russians took possession of the boat. The Russians opted not to take possession at that point and the ship was eventually scrapped.

Nordenfeldt III

# 1885-Goubet

Frenchman Claude Goubet built a battery-operated submarine in 1885. This submarine proved to be awkward and unsuitable for ocean travel. Goubet went on to build the 'Goubet II' which was also a small electric submarine, and was also mostly unsuccessful.

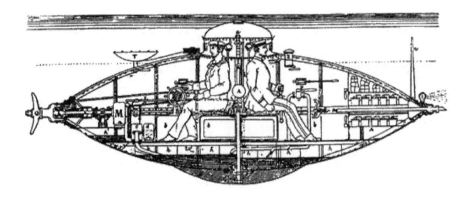

# 1897-Lake

Simon Lake was born in Pleasantville, New Jersey in 1866. As a child, Lake read Jules Verne. Lake became a mechanical engineer and began designing an underwater salvage boat in 1894. He called this boat the 'Argonaut Jr.'

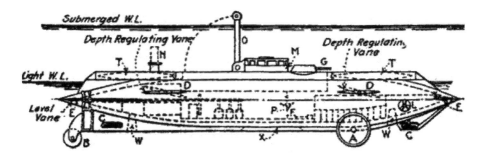

This submarine was cheaply built of two layers of pitched pine with canvas between layers. The hull was then painted with tar. It was fourteen feet long, 4.5 feet wide and 5 feet high. It had two large wooden wheels to run along the ocean bottom and a small rear wheel for steering. It was powered by a hand crank attached to a bicycle chain. This boat did have an air-lock. Lake even designed a diving suit to accompany his submarine. The helmet was made of sheet iron and the tunic was made of painted canvas. Lake lashed window weights to his legs as ballast.

After a puff-piece was written on Lake's submarine by the newspapers, the public donated money and Lake began construction on the 'Argonaut I.' Launched in 1897, this boat was 36 feet long with huge seven foot diameter cast iron wheels for running on the ocean bottom. Its power source was a thirty horse power gasoline engine with a snorkel attached to a buoy. Therefore, it had a very limited operating depth. One of the boat's unique features was a detachable keel. If there was an emergency, the keel could be dropped and the crew would be floated to the surface.

# William Carey - 1891

An Englishman named William Carey designed a diving rig using steel with ball joints used for articulation. The elongated helmet had three portholes. This diving rig would have looked like a torpedo with legs.

# Buchanan and Gordon-1894

Two Australians, John Buchanan and Alexander Gordon, devised a suit using accordion-like arms and legs.

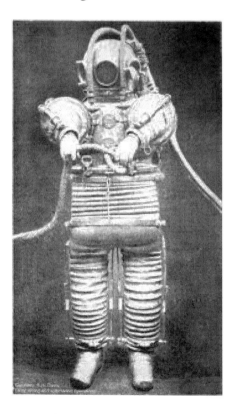

# 1898-Narval

In 1898, Maxime Laubeuf (1864-1939) built his 188 foot 'Norval.' This boat displaced 136 tons and was designed to be steam powered, but the engines were switched over to diesel.

# 1905 John Scott Haldane

John Scott Haldane, Scottish physicist, was born May 3, 1860 in Edinburgh. He attained his education at Edinburgh University and the University of Jena. His discoveries include several inventions and most notably, the first decompression tables. Many of his experiments included locking himself in a chamber under pressure or with poisonous gases to document their effects. Haldane also tested the effects of carbon monoxide by closing himself in a sealed chamber.

Haldane was the first to discover Nitrogen bubbles in the blood of saturation divers. He also visited coal mining accidents to investigate the effects and gases that had caused death. He even designed respirators for personnel. Haldane suggested using small animals in mines to detect poisonous gases. Canaries or mice worked best. Due to their faster metabolism, smaller animals react more rapidly to poisonous gases. Canaries are still used in some mines today.

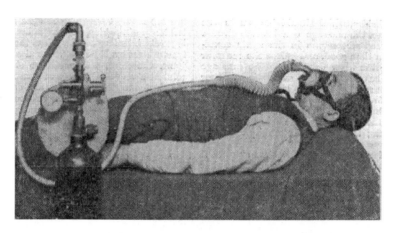

In 1907, Haldane developed a decompression apparatus which would help keep deep-sea divers safer. From his extensive experimentation and calculations, he formed the first decompression tables. These tables, though slightly modified since the invention of the computer, are still in use today.

Haldane discovered that when hemoglobin loses its oxygen, it is free to take on carbon monoxide. Inversely, when hemoglobin takes on carbon monoxide, it loses its ability to carry oxygen. This is known as the 'Haldane Effect.'

Before the start of WWI, Haldane led an expedition to Pike's Peak, Colorado to study the effects of low atmospheric pressure on the body. During WWI, when the Germans were using poison gas on their enemies, Haldane was sent to the front lines by the British government to discern the gases being used. This led to his invention of the gas mask. Gas masks have saved countless lives in fires, mining, and war.

Haldane was a Fellow of the Royal Society, a member of the Royal Society of Medicine and the Royal College of Physicians and was awarded with several accolades. Haldane changed cardiac medicine when he devised a method to measure cardiac output. He founded *The Journal of Hygiene*. Haldane become president of the English Institution of Mining Engineers. John Scott Haldane died in Oxford, England March 15, 1936 after returning from an expedition to the Middle East investigating heat stroke in oil workers.

# 1906 U-boots

In 1906, the first German U-boot was launched. U-boot is short for the German 'untersee boot,' or undersea boat. This submarine was 139 feet long and displaced 239 tons. It had a range of 2000 miles and could maintain a speed of 11 knots.

World War I began in August of 1914. On September 22$^{nd}$, 1914, a single German U-boot sank three British cruisers in just over an hour killing 1,400 sailors. During the first year of war, the German U-boots sank nearly 300 British and allied ships. The world was forced to take notice of the efficiency of these submarines.

As during all wars, new inventions were born out of necessity. More powerful diesel engines and stronger batteries were constantly being improved. Hydrophones were designed to listen for submarines underwater, so that depth charges could be properly dropped near their targets. Torpedoes had gyroscopic guidance systems. Scuba gear (self contained breathing apparatus) became much safer and more refined. Although war is often horrific, many of the inventions we take for granted today evolved from necessities of war.

# 1913 Draegerwerk, Lubeck, Germany

In 1913, Draeger produced a series of interconnected pressure simulation tanks for the research of diving and pressure-related science. The chamber was made of riveted steel and stood approximately nine feet tall and had a six foot diameter. The lower part could be filled with water to more accurately simulate dives and the chamber could achieve pressurization of about six atmospheres absolute. A bulky manhole cover-like lid was used to seal the chamber. This lid proved so cumbersome that a block and tackle was required to shift it.

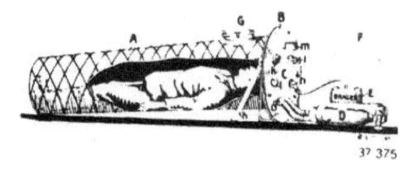

Also designed by Draeger was the first collapsible recompression chamber. This chamber was intended to be used for divers. The chamber body was made of diving suit canvas coated in polymer to make it air-tight and netted with chains to prevent rupture and bulging. The carrying case was a bulky wooden box and the door was plate steel. There was a porthole over where the diver's face would be. Pressure was achieved by a lever pump. Although it was portable, only a very low pressure could be achieved and this device was probably not very effective.

Bild A.
Chicagoer Feuerwehr mit Draeger-Halbstunden-Apparaten
Modell 1910.
(Nach einem Original-Photogramm der Chicagoer Feuerwehr.)

However, Draeger still operates out of Luebeck, Germany today with offices worldwide. Draeger holds patents on a multitude of diving and safety-related instruments and inventions. These include gas masks, scuba devices, respirators, hospital ventilators, heart monitors, anesthesia monitors, gas analysis devices, fire fighter locator devices, thermal imaging cameras, drug test kits, safety training modules and alcohol breathalyzers. If anyone is in the field of hyperbarics, they have heard the name Draeger.

# Neufeldt and Kuhnke-1913

Neufeldt and Kuhnke designed a new ball-and-socket joint using ball bearings and interlocked spheres. They built a diving rig in Hamburg in 1917 that consisted of a large metal cylinder with articulated arms and legs. It had a porthole over the face and large air cylinders on the back of the rig. Neufeldt and Kuhnke made improvements and a second rig was tested by the German Navy in 1924. A third rig was built by 1929 with even further modifications. The air tanks were covered and the hatch was now near the top of the head. The buoyancy tank was moved to the waist area. It was rated to 700 feet.

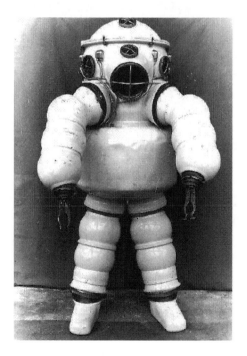

MacDuffy's diving suit design of 1914

# Harry L. Bowdoin-1915

Harry Bowdoin designed his patented diving rig using an oil-filled rotary joint design. The Bowdoin suit was never actually constructed, but would have had a front- mounted searchlight and 18 joints in the arms and legs.

# Campos-1922

Victor Campos, a Mexican-born New Yorker constructed a diving rig using oil- filled rotary joints that closely resembled Bowdoin's design. It was tested at 600 feet, but no remarks are made as to how well it functioned.

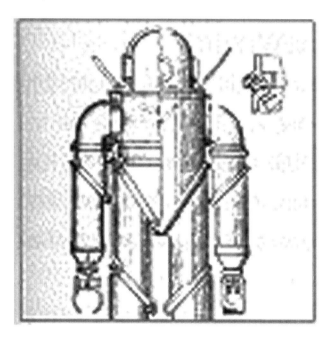

# Joseph Salim Peress - 1922

Joseph Peress, born in 1896, was a British diving engineer. He patented an oil-filled universal joint in 1922. In 1924, he constructed his first prototype diving rig out of stainless steel. But the suit proved to be too cumbersome and had several design flaws. In 1930, Peress designed a second suit of cast magnesium alloy with several modifications. It was tested to 447 feet of seawater by Jim Jarrett. This atmospheric diving suit became known as the 'Tritonia.'

The suit was again used by Jim Jarrett in 1935, when he dove on the famous wrecked ocean liner, the *Lusitania*. Peress approached the British Royal Navy with his suit, but the Royal Navy showed no interest.

Joseph Peress had designed the prototype for all modern atmospheric dive suits. However, discouraged by lack of interest in his creation, Peress went in to plastic molding and eventually formed a company. That company became the largest manufacturer of aircraft gas turbines in the world.

# 1927-Momsen Lung

Lieutenant Charles B. Momsen made a self-contained breathing apparatus so that sailors would have a way to escape more safely if trapped underwater. This apparatus resembled a gasmask with a bag attached. Momsen used a chemical filtration device to 'scrub' the $CO_2$ out of the exhaled air so it could be re-breathed.

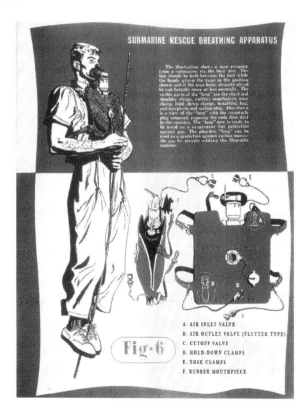

# 1928 Dr. Orval J. Cunningham

Dr. Orval J. Cunningham operated the largest hyperbaric chamber at that time in Lawrence, Kansas in 1921. The chamber was eight feet tall and thirty feet long. He used the chamber to treat Spanish Influenza. Cunningham noted a higher mortality rate in patients at higher altitudes in the Rocky Mountains as opposed to the patients in the valleys. He assumed this could be remedied with hyperbarics. Although he claimed fantastic results, due to a mechanical failure one night, a catastrophic decompression occurred and all the patients were killed.

Cunningham's large chamber in Kansas City

However, a millionaire ball-bearing manufacturer from Canton, Ohio named Henry Holiday Timken had been successfully treated by Cunningham and cured of his uremia. H.H. Timken gave Cunningham one million dollars in 1928 to operate what would later become known as the 'monster steel ball.' This giant steel hyperbaric hospital was built in Cleveland, Ohio overlooking Lake Erie. Also known as the 'Timken Tank' and the Cunningham Sanitarium, the ball stood 5 stories tall, was 64 feet in diameter, and weighed 900 tons. Each floor had twelve bedrooms. There were 38 rooms, including a smoking room on the top floor. The ball had 350 portholes. It was lavishly decorated with carpet and reading rooms and all the amenities of a hotel. In this giant chamber, patients could live up to a week under the pressure of thirty psi.

Inside the lavish hyperbaric hotel

Due to criticism from the AMA and Cunningham's complete lack of empirical evidence of its benefits, the steel ball was only in use by Cunningham for a brief period. It was sold several times and was eventually sold for $25,000 to be used as scrap metal for World War II. It is now the site of St. Joseph High School.

# 1930 - Bathysphere

Zoologist William Beebe and engineer Otis Barton worked together to create the bathysphere. Bathysphere was named from the Greek, meaning 'deep orb.' This giant steel submersible ball was made of 1¼ inch steel and weighed 5,400 pounds. It had a manhole-like entrance, two quartz viewports and two small air tanks. It was suspended by 7/8$^{th}$ inch cable and had a rubberized cable containing electrical lines and telephone lines. It also had a very strong searchlight. In 1934, the bathysphere descended to a depth of 3,028 feet of seawater. Due to Beebe's research, several new undersea life forms were discovered.

# 1930s Siebe Gorman and Company, LTD.

Augustus Siebe was born in Saxony, Germany in 1788. Siebe accepted an apprenticeship as a metalworker in Berlin. After the Battle of Waterloo, he settled in London, England.

Siebe married and went on to have nine children. In 1828 Siebe invented and patented a rotating water pump. His son-in-law was named Gorman and the Siebe Gorman Company was formed.

In 1838 George Edwards proposed safety improvements to the open diving suit designed by Charles Deane. Edwards generously gave the suggested designs to Siebe so that diving could be made safer. By 1839, Siebe began mass-producing the improved design of a closed diving suit which was fastened to the breast plate of the diving helmet by twelve bolts. Although Siebe died in 1872, this closed suit was adopted by the Royal Navy, and soon the Siebe Gorman Company became largely successful.

The company of Siebe Gorman went on to make diving helmets of copper and brass. This helmet would become the prototype for all future diving helmets. The helmets had an air inlet and an air exit. Some had an over-pressure release valve and various rounded windows.

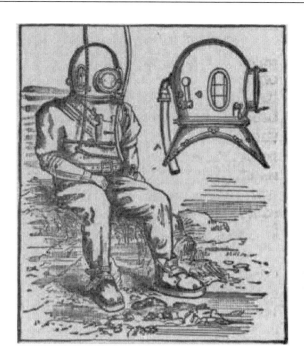

In the 1930's the Siebe Gorman Company produced a large diver recompression chamber. This large six foot diameter chamber was very well-designed. It was divided into two parts, a personnel lock and a treatment chamber. It was comprised of riveted steel plates with a large easily-accessible steel door which sealed from the internal pressure. It had oval windows, a diver telephone, electric lighting, wooden floor and a seat for the diver. At the control panel was a clock, thermometer and pressure valve. This chamber was the meat-and-potatoes version of a modern multi-place chamber.

The Siebe Gorman Company went on to produce all types of diving equipment.

This included breathing apparatus, dive suits, diving boots, pumps, knives and of course, the famous brass helmets. They also produced frogman scuba equipment for the British special forces in WWII.

The company was eventually destroyed by fire and does not exist today. However, the obituary of Augustus Siebe described him as the father of diving. His original copper and brass diving helmets are highly collectible and sell for thousands of dollars.

# 1939-Rescue Chamber

In 1939, the U.S.S. Squalus sank on a test dive trapping 33 survivors inside at a depth of 243 feet of seawater. The rescue ship 'Falcon' arrived, and onboard was the rescue chamber designed by USN Commander Allan McCann. The chamber was descended three times and all 33 sailors were saved.

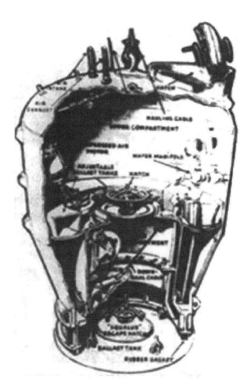

# 1948 - Bathyscaphe

Auguste Piccard was a Swiss professor of physics with an adventurous spirit. He had already made several balloon ascents into the stratosphere. Next for him would be the ocean depths. In 1948, Piccard invented and tested the 'bathyscaphe,' named from the Greek for 'deep boat.'

Photo # NH 96805   Trieste's pressure sphere, ca. 1958-59

This two-man submersible could move due to a pair of electric motors. It had a massive float made of sheet-metal filled with gasoline. This was not part of the submersible, but rather a buoyancy device. Since gasoline is lighter than water but doesn't mix with water, it provided buoyancy. Gasoline does not compress like a large tank of air would, therefore could withstand the crushing pressure of the deep. The bathyscaphe also had releasable ballast. Cousteau's group contributed a mechanical claw that could be used for gathering samples from the ocean depths. The vessel also had a seven-barreled harpoon gun. This harpoon was operated by water pressure and the harpoon tips contained stun devices. This craft had life support for the two-man crew for up to 24 hours. A radar mast was used to radio for surface support, because the bathyscaphe hatch could not be opened from the inside. A Geiger counter was also used to detect any radiation.

Photo # NH 96801 Trieste hoisted out of water, circa 1958-59

Bathyscaphe with massive gasoline float

After the French ignored Piccard's recommendations for improvements, Piccard raised private funds to build another bathyscaphe, to be called the 'Trieste,' after where it was to be built. The Trieste project was taken over by Cousteau's group.

The Trieste was a 6 foot 7 inch sphere underneath a gigantic float filled with 25,000 gallons of gasoline. It had nine tons of steel shot for ballast and two water tanks that were flooded to submerge. It, too, had electric motors and the windows were cone-shaped 6 inch thick plexiglass. In 1954, the Trieste achieved a depth of 13,284 feet of seawater. Many useful experiments were performed, including studies on light waves, sound waves, salinity levels and new life forms.

# 1950's Professor Doctor Ite Boerema

Ite Boerema was born in Holland October 14th, 1902. He came from a poor family, the son of a river boat captain. Boerema went on to Groningen University where he succeeded academically. During World War II, Boerema played an active roll in the Dutch underground resistance by smuggling people out of the hospital. He was twice taken prisoner by the Germans, but due to an extreme shortage of doctors, he was released. Later, he would be decorated for his service.

Boerema went to America where he performed the first heart and spleen transplants on animals. By this time he had attained his M.D. and Ph. D. and had studied orthopedic surgery, neurosurgery, cardiovascular surgery and was now paving the way for future transplantations.

In 1946 Boerema returned to Holland and became professor of surgery at the University of Amsterdam. In 1955 he went to the military installation Den Helder to use their hyperbaric chamber to study benefits of hyperbarics in cardiac surgery. Here, he found that cardiac arrest could be induced for approximately eight minutes under hyperbaric conditions without irreversible damage as opposed to roughly four minutes in a normobaric environment.

Boerema began experiments on animals. One of his most famous findings was that hyperbarics could force oxygen not only into the hemoglobin, but also into the blood plasma. Therefore, he preformed an experiment in which he removed all the blood cells from a pig, leaving only the plasma. The pig lived due to the high amount of oxygen dissolved into the pigs' plasma. After the experiment was done, he reintroduced the blood cells to the pigs' vasculature and the pig was fine. He then published *Life Without Blood* documenting his experiment.

In 1959, the manufacturer van Leer and the city of Amsterdam donated a hyperbaric chamber to the Wilhelmina Gasthuis Hospital for Boerema's research. In 1960, he successfully treated gas gangrene, a deadly bacterial infection that devours tissue rapidly. December 18th, 1960, Boerema performed the first open heart surgery in a hyperbaric chamber. He also treated carbon monoxide poisoning and vascular complications successfully with hyperbarics. These diagnoses are still treated with hyperbarics and covered by most insurance carriers today.

In his life, Boerema published more than 250 papers. He was decorated for his service during WWII, and honored with several titles and awards such as the Knight of the Order of the Dutch Lion, Officer of the Legion of Honour, member of the International College of Surgeons, the American College of Surgeons, and the British and American Thoracic Surgery Associations.

# 1955-U.S.S. Nautilus

The world's first nuclear submarine, the 'Nautilus,' was built in 1955. It crossed under the North Pole in 1958. The water temperature was a mere 32.4 degrees Fahrenheit.

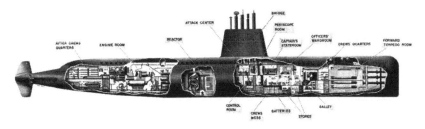

# 1957-Conshelf Habitats

Navy Captain and Doctor George Bond considered saturation diving habitats in 1957. This would enable a diver to stay or live underwater until work was completed. Navy superiors, however, were less than enthusiastic. Oceanologist Jacques Cousteau and inventor Edwin A. Link got wind of the idea and ran with it.

Jacques Cousteau

Jacques Cousteau set up a habitat in the Mediterranean near Marseilles. It was named 'Conshelf I' after the continental shelf it sat upon. This 8 foot by 17 foot structure sat at a depth of 33 feet of seawater, which is two atmospheres absolute. This habitat was linked to a surface ship for electricity and phone lines. Two divers spent a week in the habitat with no apparent side effects.

Cousteau set up Conshelf II in 1963 near the African Coast of the Red Sea. This habitat was larger than Conshelf I and could house five divers for thirty days. The main house of Conshelf II was dubbed 'Starfish House' due to its unique shape. This habitat rested at 36 feet of seawater with an additional area called 'deep cabin' positioned at ninety feet of seawater. Two divers could spend a week in the deep cabin. There was also an underwater hangar for the 'diving saucer,' a submersible that could be utilized to 1,000 feet of seawater.

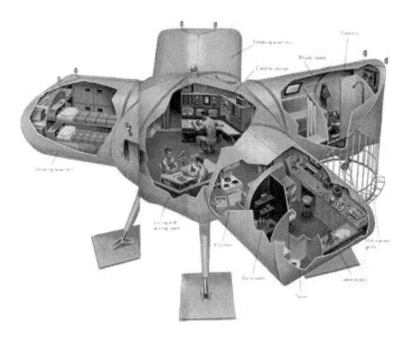

Conshelf II

deep cabin

Conshelf III was built in the Mediterranean in 1965. It was much deeper than the previous two habitats, resting at 328 feet of seawater. This was an 18 foot sphere divided into two stories. The lower story was used for sleep, diving operations and bathroom space. The upper story space was used as an office for communications, data collation and dining. Conshelf III could house five men for two weeks.

# 1964-Sealab

In 1964, the U.S. Navy constructed Sealab I near Bermuda under the leadership of Doctor Bond. This habitat could house four 'aquanauts' for two weeks and laid at a depth of 193 feet of seawater. The steel structure was forty feet long with a diameter of ten feet and resembled a fat cigar. Supplies such as newspapers and food were sent down daily in sealed pressure containers.

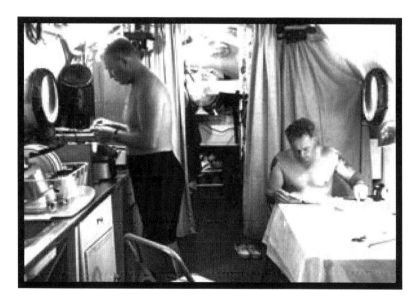

Inside Sealab I

Soon after, Sealab II was built off of La Jolla, California at a depth of 205 feet of seawater. This 57 foot long cylinder was constantly monitored by the use of closed- circuit television cameras. It was dubbed the "Tiltin' Hilton" because it rested on a sloped ledge on the sea floor.

Sealab II

# 1964 Draegerwerk, Germany

As hyperbaric chambers were being developed all over the world, many were designed poorly and with very little knowledge or consideration for safety. For example, many still had carpeting or a complete lack of backup systems for air, oxygen, or power.

In 1964, Draeger designed what could be considered the first monoplace chamber. Although the chamber was very basic in design, it is the same general design that is used in most monoplace facilities today. It was a cylindrical tube approximately 7 feet long with a 28 inch diameter. It was set on a framework placing the chamber at waist level. The patient would lay on a gurney and could be slid into the chamber for easy access. Oxygen was inspired via a regulator mask. The chamber had an intercom and porthole windows located above the patients' head. This unit was on wheels and could easily be moved or repositioned.

# 1967 Wurzburg, Germany

In 1967, Draeger manufactured a large multiplace hyperbaric chamber for the University Hospital of Wurzburg, Germany. The chamber was approximately 6 feet in diameter and could accommodate a stretcher or several patients on benches. Masks were used for the inspiration of oxygen. The door was large and rectangular, as are the doors on most modern multiplace chambers still today. Draeger advertised that this chamber could effectively treat cardiac defects, burns, carbon monoxide poisoning, and arterial or venous insufficiency.

As for hyperbarics, technology and safety were now being considered. This era brought true scientific research and thorough documentation of case studies. In fact, the medical conditions that this Draeger chamber claimed to help are now covered by most insurance carriers as legitimate treatments.

# 1970 Draeger Oxyfulm Monoplace Chamber

The theory was, by hyperoxygenating cancer cells, they would become more susceptible to radiation. Therefore, a chamber was designed that could deliver hyperbaric oxygen while the patient received radiation.

Draeger two-man hyperbaric chamber

Draeger created a monoplace chamber which was cylindrical in shape and made of inch-thick acrylic. The chamber could be readily moved into the field of radiation. This chamber also had several door penetrators. Acrylic hulls and door penetrators are common features found in todays industry-standard chambers.

Draeger hard suit still in use today

# JIM Suit-1969

Two workers of Underwater Marine Engineering, Limited, otherwise known as UMEL, shared ideas on creating a practical atmospheric diving suit. These two co-workers were Mike Humphrey and Mike Barrow. They contacted Peress and acquired the Peress Tritonia suit from a junk shop. The UMEL team received a grant from the British government, then set up the company DHB to carry out the research on the diving suit project.

DHB (Dennison, Hibberd, Barrow) designed a new atmospheric diving suit using the basic Peress joint idea of an oil-supported universal joint. They designated the new suit "JIM" after Jim Jarrett, the test-diver for the Tritonia. The "JIM" suit was tested at the HMS Vernon dive tank.

This JIM suit had multi-functional claws or manipulators to accomplish tasks and work with tools. It had 6 acrylic porthole windows and a top hatch. The suit stood six feet six inches tall and weighed approximately 1000 lbs. Its life support system consisted of a face mask with a $CO_2$ scrubber and two 800 liter oxygen bottles carried externally and could support a diver for twenty hours. The magnesium alloy body also had an ultrasonic pinger and a strobe light and the ballast could be jettisoned to surface since the suit had a negative buoyancy without ballast.

# WASP Suit-Mid 1970's

Graham S. Hawkes wanted to design a suit using propulsion so that the diver could operate and move more freely in water, since the JIM suit could not operate in a sea current. The WASP suit was developed and was basically a JIM suit with thrusters and no articulated legs. The hull was made of cast aluminum alloy and the trunk was made of resin-impregnated fiberglass an inch thick. Four electric thrusters could be controlled with foot pedals and the diver had a 360 degree field of vision through the 16mm thick glass dome. This atmospheric diving suit was tested to a working depth of 2,000 feet of sea water and could travel up to two knots.

# 1983 Hyox, Scotland

In 1983, the Hyox Company, formerly Airox Hyseal, designed a monoplace chamber with safety and ergonomic aspects considered. The chamber had an externally sleek, modern appearance. This chamber had a telescopic reclining chair that was easily slid into the chamber. The door was stowed on the inside bottom of the chamber and slid upwards to the frame. The control panel was located conveniently on the side of the chamber and there were three well-located windows that enabled the patient to view television. The chamber would be compressed with room air and the patient would inspire oxygen via a mask. The patient could even depressurize the chamber from the inside. In the medical field, this could be seen as a blessing or a curse.

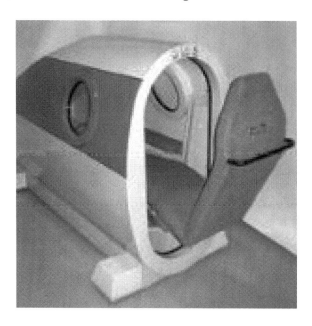

# NEWT Suit-1984

In 1984, Phil Nuytten patented an oil-filled rotary joint. Nuytten designed the NEWT suit to be used in the oil industry. This suit, having 20 joints, increased mobility and more closely hugged the human form. It was made of aluminum and weighed 1,100 lbs. The life support system could support a diver for forty hours.

# Perry Baromedical Corporation

Perry Baromedical Corporation has a very interesting history. John H. Perry, Jr., publisher of 28 newspapers in Florida and the Bahamas, formed a 3-man company called Perry Submarine Builders in 1956. This small company built the Perry Cubmarine, small submersibles for a one or two-man crew. One such submarine was used in the search for a hydrogen bomb that was lost when a B-52 crashed in 1966. In 1965, Perry joined forces with Edwin Link, the inventor of the Link Trainer, a device used to train WWII pilots. In 1968, the company built its first diver decompression chambers for the oil industry. By 1969, the company was building saturation diving systems, remote control underwater vehicles and submersibles. Interestingly, the company built the submersible Lotus sports car used in the 1977 James Bond film, *The Spy Who Loved Me*.

In 1979, a division was formed called Perry Baromedical that was specifically devoted to making hyperbaric oxygen chambers for the medical field. In 1986 came the Perry Sigma II, the world's first acrylic multi-place chamber. By 1989, Perry had made its first mono-place chamber, the Sigma I. In 1996, Perry made the Sigma Plus, a large and very comfortable mono-place chamber. This chamber has a 40 inch internal diameter, an acrylic cylinder that can be pressurized to 3 ATA with oxygen or air and a sleek, futuristic design that is aesthetically pleasing to the eye. It can be fitted with television, DVD player and radio. One of the highlights of the chamber is its large innerspace, enabling a patient to sit upright or recline.

Perry is currently the only company that produces mono-place, bi-place, and multi-place chambers of the highest quality.

# Sechrist-Anaheim, California

Sechrist Industries, founded by Ron Sechrist in 1973, has created state-of-the-art chambers that have become today's industry standard. In 1979, Sechrist manufactured the first infant ventilator controlled by a microprocessor.

In 1995, Sechrist launched the model 3200. Since this chamber is highly preferred in the modern medical field, this author will go into greater detail. The Sechrist 3200 has a 32 inch diameter which enables patients of average size to move freely inside the chamber. Bariatric models are also available. The chamber is capable of 3 ATA, which is equivalent to a pressure of 29.4 psig or 66 feet of sea water. The cylinder is made of nearly inch-thick acrylic, allowing the patient to see in virtually any direction, thus decreasing any feelings of claustrophobia. Bear in mind that before this chamber, most chambers had no windows or only a porthole above the patients' face, which could be very intimidating to the patient. This also enables the chamber operator to view the patient without the use of cameras. It can be fitted with tailor-made flat screen televisions and DVD players for patient entertainment. The vaulted door has several access points, called penetrators, that allow cables and IV tubing to pass through the door if patient monitoring should be necessary. The door also has a pin mechanism for safety that prevents the chamber from being opened while under pressure. The patient is grounded at all times via an elastic wrist-strap which is grounded to the chamber hull. It is capable of compressing with room air or oxygen, although in most cases, oxygen is used as the compressing gas. A penetrator for a mask used to facilitate air breaks is also available. Air breaks reduce the risk of oxygen toxicity, especially during longer or deeper treatments. The carts have been modified to accommodate a tray with a mattress that slides the patient effortlessly into the chamber on rails. The workings of the chamber operate solely on pneumatic pressure, therefore if a power outage should occur, the chamber could still be operated safely. There is an intercom system for two-way communication with a battery back-up. The control panel also has a dial for adjusting the rate of compression/decompression. There is also an emergency vent button if the chamber should need to be emergently decompressed. There are rumors that a water deluge system could soon be included for further fire safety. The chamber has two over-pressure release valves for added safety. This chamber weighs just under a ton, but can be moved as it has wheels which can be hidden by metal cowling.

Sechrist 3200

This chamber is extremely well-designed. Sechrist strives for safety and simplicity while maintaining a sleek attractive appearance. Since Sechrist manufactures their own machines, they can guarantee the quality control of each piece of material that goes in to the construction of their chambers.

Sechrist, located in Anaheim, California, will continue to set the standard by incorporating new ideas and the latest research. Safety, functionality and innovation have been combined to produce the finest monoplace hyperbaric chambers in the medical arena.

The newly-installed second largest hyperbaric chamber at St. Luke's Medical Center, Milwaukee

# HBO Contra-Indicated Items Approved for Dive in a Monoplace Chamber

**(Note: items approved for a mulitplace chamber may vary)**

Note from author: This list of items that MAY AT TIMES be allowed in a monoplace hyperbaric oxygen chamber is in no way meant to be adopted as policy and does not reflect the opinions of any health community or health care facility. This list should be used as a guide. Always review these or any items to be taken into a hyperbaric chamber with your safety director and hyperbaric officials at your facility! Some or all of these items may be deemed unacceptable to take into a chamber at your facility.

EYEGLASSES-all eyeglasses may be taken into the chamber. Use good judgment as to the mental stability of the patient. Titanium frames do have a higher stress point, which means if one were to intentionally bend them back and forth until they broke, the friction heat generated could ignite the oxygenated materials in the chamber.

MEDICATION PATCHES-Do not allow nitroglycerin patches or pastes, aluminum or mylar patches, solely due to the metal content, or patches with air bubbles such as Fentanyl, as they may rupture or the rate of absorption may be altered. For all other patches, use good judgment and ask your safety director before final approval.

DRESSINGS-Dressings should be considered on a case-by-case basis. All petroleum dressings may be occluded by gauze or a moist washcloth. It is better to occlude a dressing than to take down a dressing or to let microbes leak out onto the HBO cart. Just because the word petroleum is in the name of the dressing does not mean it has been dipped in gasoline.

CLOTHING-Technically, we could allow in anything that says 100% pure cotton. But for safety, one should allow only specially washed and specifically manufactured all-cotton scrubs, gowns and blankets to limit other materials sneaking into the chamber. Synthetic fabrics have a higher tendency to build up static charge and should be strictly limited.

BOOKS AND MAGAZINES-Many people think it is ok to have a harmless book, which in theory, it is. But, it is one of the fastest burning fuels in a pure oxygen environment. Therefore, no books or magazines should be allowed into a monoplace chamber. Magazines often have petroleum-based inks. Burn a magazine sometime. It will burn with a blue-green flame.

SYRINGES and WATER BOTTLES-Although patient comfort is important, we always want to limit the amount of burnable items in the chamber. But if you have a patient that insists on having a drink nearby, then use these rules. Water only unless a non- carbonated juice is needed to maintain blood glucose levels. The container MUST be vented. That is, if it is a syringe, it must not be capped or under pressure. Water bottles must be pure plastic with easily identifiable one-piece molded parts. Make sure these are also vented. No air bubbles or advertising inserts or insulated bottles with chrome lips should be used. Do not use lined or double-walled water bottles. Keep in mind that the chamber operator is responsible for spills.

CATHETER BAGS-Even though they have a big metal clip, they are probably alright to take in the chamber if cleared by your safety director. We're not putting these things in the chamber with big hunks of flint. Metal doesn't spontaneously combust and catheters are more important to take in than say . . . car keys. So we will make allowances. But unless you want to soak up urine after the dive, drain the bag pre-dive and try to burp the air out. The less air left in the bag, the less it will expand during decompression. Theoretically, if we dive to 2 ATA, the bag could be half-full of air and still not burst.

ORTHOTIC FIXATORS, ORIF BRACES-These giant hunks of metal, although made of titanium, may go in the chamber. If you are strong enough to bend those titanium bolts back and forth until it breaks, then you could probably bust through the acrylic and save yourself. HOWEVER, you must wrap it with a towel to protect the acrylic from scratches. A scratch of more than

0.1mm could cost $55,000 to replace the acrylic. So if the patient is confused or spastic, you might want to ban the ORIF device on that occasion.

CASTS-Although we would prefer plaster casts, most casts are now a fiberglass material. NOT COTTON, however, one can allow the cast in the chamber. Make sure the cast has cured at least 12 hours. As an extra precaution, one could place a dampened towel around the cast, because sparks cannot survive at 55% humidity or more. Again, watch the acrylics. NO AIR-CASTS!

PAPER TAPE is ok to use to tape towels in place or to occlude dressings. Paper is flammable, but paper tape is usually used in small amounts. It is not optimal, but it is better than using plastic tape or silk tape.

DENTURES-One may inspect the dentures first, but the odds of dentures sparking in such a humid environment are so low, that most facilities will allow them entry. Having teeth also aids the air-break mask to fit more firmly to one's face.

HEARING AIDS-Nothing with hidden internal parts that you cant inspect should be allowed into a hyperbaric chamber. Plus, hearing aids have batteries and electronics. NEVER allow batteries or electronics in the chamber. If a patient cant hear, use an alternate method of communication, even if the patient has to be taught hand signals. Dry erase boards are an effective method for communication from outside of the chambers. This does not include internal devices such as pacemakers. Each model of pacemaker should be individually cleared for hyperbaric chambers. Most modern pacemakers have already undergone extensive testing by the manufacturer. The manufacturer may need to be called directly by the safety director. Most modern pacemakers have been tested to 6 ATA.

WEDDING RINGS-Some hyperbaric patients haven't had their rings off in 30 years. Many of their wedding rings would have to be cut off. This is a rule we can bend a little. No diamonds should be allowed in the chamber because they could scratch the acrylic or the HBO mattress. No two rings that could touch each other should be allowed in if avoidable. BUT, one wedding band is acceptable and the operator could occlude it with PAPER TAPE! Marital bliss!

CANDY/CHEWING GUM-Although not a direct threat to the chamber, candy and gum are choking hazards, especially while lying flat. If a patient were to choke, the operator would generally not be able to help the patient for several minutes.

In closing, use good judgment. Always ask the safety officer if you are not sure. If one crosses something that hasn't been researched, find an answer or ban the item until your safety director finds the correct information. I tried to address some of the items that are not on the standard lists, so that there would be less confusion later. I tried as well to explain the logic behind each of the no-no's or why one might allow certain banned items into a monoplace chamber.

# Items not approved to be taken into a monoplace chamber

Shoes
Wallets Money/coins
Hand warmers
Electric blankets
Watches
Nail polish
Deodorant
Cologne/perfume
Make-up
Stockings/nylons
Solar devices
Ink pens
Glucometers
Chewing gum
Thermos jugs
Sharp objects
Socks/underwear
Pictures/photos
Alcohol pads
Combs/brushes
Adult briefs/diapers
Street clothing
Cell phones
Electronic devices
Video games
Hair spray

**Hair gel**
**Lotions**
**Oils**
**Newspapers**
**Books**
**Magazines**
**Medicines**
**Jewelry**
**Cigarettes**
**Lighters or matches**
**Batteries**
**Metal objects**
**Bobby pins/barrettes**
**Hair clips/scrunchies**
**Pagers**
**Car keys**
**Thermometers**
**Ear plugs**
**Head phones**
**Hard contact lenses**
**Hot water bottles**
**Soda cans**
**Artificial limbs**
**Knitting/sewing**
**Playing cards**

# Glossary

Absolute pressure—the sum of all pressures, both atmospheric and hydrostatic, exerted on a unit of area.
Acoustic torpedo—torpedo guided by sound.
Aft—pertaining to the rear or stern of a ship.
Air compressor—machine that raises pressure to greater than one atmosphere.
Air embolism—gas bubbles in the blood stream from burst lung tissue.
Air lock—a double hatch allowing access to a pressure vessel or submarine while maintaining internal pressure.
Ascent—movement toward reduced pressure or toward the seas' surface.
ATA—(atmospheres absolute) the sum of air pressure and water pressure.
ATM—(atmosphere) pressure the Earths' atmosphere exerts on a body.
Aqualung—self-contained underwater breathing apparatus.
Aquanaut—Navy term for a diver working at depth or in a habitat.
Attitude—position of a submarine in water.
Ballast tank—tank used to adjust buoyancy and trim in a submerged vessel.
Barodontalgia—tooth pain related to changes in pressure.
Barotrauma—tissue damage related to changes in pressure.
Bearing—compass direction of an object from the observer.
Bends—medical condition caused by formation of gas bubbles in blood vessels. See decompression sickness.
Berth—place where a ship is anchored.
Blockade—where naval ships cut off access to a harbor or waterway.
Blow—to expel water from a ballast tank by using the force of compressed air.
Bow—pertaining to the front of a ship.
Boyle's Law—If the temperature is constant, the pressure of a gas will change inversely with the volume.

Bridge—structure on or at the top of a submarine containing controls and optical capabilities.

Caisson—a pressurized area used to force water out of an underground construction site.

Caisson sickness—decompression sickness acquired while working in a caisson.

Chamber attendant—staff member to assist a patient or occupant of a hyperbaric chamber.

Charles' Law—if the pressure is constant, the temperature varies with the volume.

Chokes—a sign of decompression sickness where patient may exhibit signs of choking such as shortness of breath, gasping or labored breathing.

Clinometer-measures roll aboard a submarine.

Compressed air—gas stored under pressure in tanks for breathing or for pressure-related needs such as purging ballast tanks.

Compression—increasing pressure by adding gas or by descending underwater.

Conshelf—Jacques Cousteau's undersea habitat named after the continental shelf for which it sat.

Contact mine—underwater mine designed to detonate when a ship makes contact with the mine.

Dalton's Law—pressure exerted by a certain gas in a mixture of gases is the same as that gas would exert alone.

DDC—deck decompression chamber.

Decompression—when pressure is being decreased or when a diver is ascending up through the water toward surface.

Decompression sickness—manifestations caused by rapid reduction in pressure, most commonly associated with a diver coming to surface rapidly.

Depth charge—high explosive charge dropped from ships or aircraft to destroy submarines.

Depth gauge—meter used to determine depth underwater.

Dive—session of exposure to increased pressure in a hyperbaric chamber or underwater.

Diving bell—a hollow bell-shaped vessel containing breathable air enabling safe descent for a diver.

Dive plane—controllable rudder or fin-like surface to steer a submarine underwater.

Dyspnea—shortness of breath.

Ear squeeze—pain inflicted on middle ear related to pressure changes.

Echo—a reflected acoustic signal.

Edema—swelling caused by fluid in body tissues.

Escape hatch—emergency exit on a submarine that may be able to open from either side.

Fathom—measure of depth equal to six feet.

Frenzel maneuver—a technique to equalize pressure in the middle ear by clamping the nose, opening the mouth and moving the tongue forward and backward.

Frogman—Navy SEAL or underwater demolition expert.

FSW—feet of seawater.

Galley—ships' kitchen.

Guy Lussac's Law—if the volume is constant, the pressure changes with the temperature.

General gas law—if any variable is changed (pressure, temperature or volume), there must be changes in the other variables. I.e., if pressure increases, the temperature must rise or the volume must change.

Habitat—an underwater structure where divers can live for extended periods of time.

Hand—crewmember of a ship.

Heliox—a breathable mixture of helium and oxygen often used in deeper dives.

Hookah—breathing apparatus on the outside of a habitat or ship enabling external divers to breathe.

Hull—body of ship or submarine.

Hyperbaric—pertaining to pressure greater than one atmosphere.

Hyperbaric chamber—vessel designed to create environment with pressure greater than one atmosphere.

Hypoxia—condition where tissue has below-normal levels of oxygenation.

Inclinometer—measures the roll of a submarine.

Inert gas—a random stable noble gas.

Keel—the longitudinal beam or the bottom of a ship.

Knot—unit of measure equal to a nautical mile; 6,080 feet.

Leeward—direction waves are traveling; away from the wind.

Log—where ship's speed and position are recorded. mmHg—millimeters of Mercury; a measurement of pressure; 760mmHg is equal to one atmosphere.

Navigator—officer designated to plot a safe course for a ship.

NFPA—National Fire Protection Association

Nitrogen narcosis—a state of dysphoria caused by the narcotic effects of Nitrogen at pressure.

Oxygen toxicity—the poisonous effects of breathing oxygen at higher pressures.

Paul Bert effect—convulsions caused by the poisonous effects of oxygen at higher pressures, discovered by Paul Bert.

Penetrator—Plug or outlet on the outer hyperbaric chamber door that allows cables, hoses and intravenous tubing to be passed through the door while maintaining pressure inside of the chamber.
Periscope—submarine device using mirrors for visualizing surface.
Ping—acoustic signal sent out by an underwater transducer.
Pitch—rise or fall of a ship in seawater.
Pneumothorax—gas trapped in the chest cavity outside of the lungs.
psi—pounds per square inch; 14.7 psi equals one atmosphere.
Pulmonary barotrauma—lung damage due to pressure changes.
PVHO—Pressure Vessels for Human Occupancy
RADAR—Radio Detection And Ranging; device used to detect vessels by radio echoes.
Rescue chamber—vehicle used to safely remove submariners from a disabled submarine.
Salvage—to reclaim materials lost to the sea.
Sand hog—worker from a pressurized caisson or tunnel.
Saturation diver—diver who lives and works in a pressurized underwater chamber.
Screw—propeller of a ship.
SCUBA—Self-Contained Underwater Breathing Apparatus.
Scuttle—to sink or disable a ship with the intention of making it unusable to the enemy.
SEAL—naval commando specialized for Sea, Air and Land operations.
Silent running—condition aboard a submarine when engines and machinery are silenced to evade acoustic detection.
Sinus squeeze—sinus pain caused by pressure changes.
Snorkel—a tube used by a diver or a submarine to draw surface air while submerged. ss—submarine.
Submariner—sailor assigned duties as a crewmember of a submarine.
Topside—slang referring to the surface or one ATA.
Torpedo—self-propelled underwater weapon containing high explosives.
Toynbee maneuver—a technique to equalize pressure in the middle ear by pinching nostrils, closing mouth and swallowing.
Treatment depth—the prescribed therapeutic pressure placed on a hyperbaric chamber patient.
U-boot—German designation for 'untersea boot,' or undersea boat. A German submarine.
Valsalva maneuver—a technique to equalize pressure in the middle ear by pinching nostrils, closing mouth and forcing air through the Eustachian tubes.

Weigh—to hoist the anchor.
Weight belt—a belt with attached ballast, usually lead weights, used to obtain desired buoyancy.
Yaw—horizontal angular rotation of a ship.

# Bibliography

Butterfield, H. *The Origins of Modern Science*. London: Bell and Sons, Ltd., 1957.

Cable, Frank. *The Birth and Development of the Modern Submarine*. New York: Harper & Bros., 1924.

Compton-Hall, Richard. *The Submarine Pioneers*. Gloucestershire: Sutton Publishing, Ltd., 1999.

"Cornelius Drebbel." Brett McLaughlin. *Dutch Submarines*. 17 Nov 2010. *www.dutchsubmarines.com/specials/special_drebbel.htm*

De Latil, Pierre. *Man and the Underwater World*. London: Jarrolds, 1956.

"De Son 1654." Cpt. Brayton Harris. *World Submarine History Timeline*. 3 Mar 2010. *www.submarine-history.com/NOVAone.htm#1634*

"De Villeroi." *World Submarine History Timeline*. 16 feb 2010. *www.submarine-history.com/NOVAone.htm#1797*

"Draegerwerk." *Draeger*. 13 Jun 2010. *www.draeger.com/US/en_US/company/about_draeger/history/*

Dugan, James. *Man Under the Sea*. New York: Harper & Bros., 1956.

"Fleuss." *Rebreathers Diving Center*. 13 May 2010. *www.rebreathers.eu/cms_rebreathers/en/node/86*

"Giovanni Alfonso Borelli." *Wikipedia*. 18 Sep 2010. *http://en.wikipedia.org/wiki/Giovanni_Alfonso_Borelli*

Harris, Gary. *Iron Suit-The History of the Atmospheric Diving Suit*. Flagstaff: Best Publishing Co., 1994.

Haux, Gerhard. *History of Hyperbaric Chambers*. Flagstaff: Best Publishing Co., 2000.

"History of the Submarine-David Bushnell 1742-1824." Bellis, Mary. *About.com*. 8 Apr 2010. *http://inventors.about.com/od/sstartinventions/a/submarines_3.htm*

"History Overview of Perry Baromedical." *Perry Baromedical*. 8 Apr 2010. *www.perrybaromedical.com/company-history.html*

Horton, Edward. *The Illustrated History of the Submarine*. Sidgewick & Jackson, 1974.

"Ite Boerema-surgeon and engineer with a double-Dutch legacy to medical technology." *SciVerse*. Elsevier, Inc. Leopardi, Metcalfe, Forde, Maddern. 20 Jun 2010. *www.sciencedirect.com/science?_ob=ArticleURL&_udi=B6WXC*

"John Scott Haldane." *Faqs.org*. 21 Aug 2010. *www.faqs.org/health/bios/55/John-Scott-Haldane.html*

"Jules Verne." Petri Liukkonen. 11 Mar 2010. *www.kirjastu.sci.fi/verne.htm*

"Lavoisier, Antoine (1743-1794)." Wolfram Research. *Eric Weisstein's World of Biography*. 7 Aug 2010. *http://scienceworld.wolfram.com/biography/Lavoisier.html*

Limburg, Peter and Sweeney, James. *Vessels for Underwater Exploration*. New York: Crown Publishers, Inc., 1973.

"Lodner D Phillips-1856." *The Rebreather Site*. 6 Jun 2010. <*www.therebreathersite.nl/12_Atmospheric%20Diving%20Suits/1856_Phillips.htm*>

"The Momsen Lung." *Louis L'Amour-The Adventure Stories*. 28 Oct 2010. *www.louislamourgreatadventure.com/PongaJimMerchantMarine11.htm*

"Important Events in Ocean Engineering History." Dr. Stephen Wood, P.E. DMES-Ocean Engineering. 12 Jun 2010. *http://my.fit.edu/~swood/index.htm*

Neubauer M.D., Richard and Walker, DPM, Morton. *Hyperbaric Oxygen Therapy*. New York: Avery, 1998.

"Newt Suit." *Diving Heritage*. 13 Jun 2010. *www.divingheritage.com/newtkern.htm*

Partington, J.R. *A Short History of Chemistry*. 3rd Ed., 1957.

"Paul Bert." *Faqs.org*. 3 Jul 2010. *www.faqs.org/health/bios/29/Paul-Bert.html*

"Questions and Answers About the Monster Steel Ball." Lakeside Press. 13 Apr 2010. *www.lakesidepress.com/pulmonary/hyperbaric/steelball.htm*

"Robert Fulton." *Wikipedia*. 16 May 2010. *http://en.wikipedia.org/wiki/Robert_Fulton*

"Robert Fulton." *World Submarine History Timeline*. 14 Feb 2010. *www.submarine-history.com/NOVAone.htm#1797*

"Sechrist Hyperbaric." *Sechrist*. 12 Mar 2010. *www.sechristusa.com/our-history.html*

Tall, Jeffrey. *Submarines and Deep-Sea Vehicles*. San Diego: Thunder Bay Press, 2002.

"William Bourne." *Wikipedia*. 18 Feb 2010. *http://wikipedia.org/wiki/William_Bourne_(mathematician)*

Workman, Wilbur. *Hyperbaric Facility Safety: A Practical Guide*. Flagstaff: Best Publishing Co., 2010.